Learning By Example Using VHDL
– Advanced Digital Design Using a NEXYS 2™ FPGA Board

Richard E. Haskell
Darrin M. Hanna

Oakland University, Rochester, Michigan

LBE Books
Rochester, MI

Copyright 2008 by LBE Books, LLC. All rights reserved.

ISBN 978-0-9801337-4-5

Second Printing, 2009

Published by LBE Books, LLC
360 South Street
Suite 202
Rochester, MI 48307

www.lbebooks.com

Preface

A major revolution in digital design has taken place over the past decade. Field programmable gate arrays (FPGAs) can now contain over a million equivalent logic gates and tens of thousands of flip-flops. This means that it is not possible to use traditional methods of logic design involving the drawing of logic diagrams when the digital circuit may contain thousands of gates. The reality is that today digital systems are designed by writing software in the form of hardware description languages (HDLs). The most common HDLs used today are VHDL and Verilog. Both are in widespread use. When using these hardware description languages the designer typically describes the *behavior* of the logic circuit rather than writing traditional Boolean logic equations. Computer-aided design tools are used to both *simulate* the VHDL or Verilog design and to *synthesize* the design to actual hardware.

This book assumes some previous knowledge of basic digital logic and VHDL as is covered in our book *Learning By Example Using VHDL – Basic Digital Design Using a BASYS FPGA board* (L-BASYS). That book, which is available from www.lbebooks.com contains over sixty VHDL examples covering basic digital design of both combinational and sequential circuits. All of the examples in that book were simulated using the Aldec Active-HDL simulator. We will use the same simulator in this book. A free student version of this simulator is available from http://www.aldec.com/Downloads/default.aspx. All of the examples in the L-BASYS book that run on the Digilent BASYS board will run on the Nexys-2 board by using the file *nexys2.ucf* (available from www.lbebooks.com) instead of *basys.ucf* to define the pin numbers.

The emphasis in this book is how to implement algorithms in an FPGA using VHDL. We will illustrate the methods by developing working programs that use all of the main features of the Nexys-2 board including the VGA port, the serial port, the PS/2 port, the onboard 16 Mbytes of RAM, and the onboard 16 Mbytes of flash memory. In addition to developing many examples of special-purpose processors we will illustrate the development of a general-purpose processor by designing a Forth core for the FPGA. You will be able to write high-level Forth code, compile it to a VHDL ROM, and execute it on the Forth core that we design in this book.

Many colleagues, students and reviewers have influenced the development of this book. Their stimulating discussions, probing questions, and critical comments are greatly appreciated.

Richard E. Haskell
Darrin M. Hanna
Oakland University
Rochester, Michigan 48309

Learning By Example Using VHDL
-- Advanced Digital Design Using a NEXYS 2™ FPGA Board

Table of Contents

1. VHDL Building Blocks — 1
   1.1 Modeling Digital Circuits Using VHDL — 1
   1.2 VHDL *if* Statement — 4
      Example 1 – 2-to-1 Multiplexer — 5
      Example 2 – Registers — 8
      Example 3 – Debounce Pushbuttons — 10
      Example 4 – Clock Pulse — 11
      Example 5 – Counters — 13
      Example 6—Clock Divider — 14
      Example 7 – Comparators — 16
   1.3 VHDL *case* Statement — 18
      Example 8 – 4-to-1 Multiplexer — 18
      Example 9 – 7-Segment Decoder — 20
      Example 10 – Arithmetic Logic Unit (ALU) — 22
   1.4 VHDL *for* Loops — 25
      Example 11 – 4-Input Gates — 25
      Example 12 – Binary-to-BCD Converter — 27
      Example 13 – Gray Code Converters — 31
      Example 14 – Multiplier — 34
      Example 15 – Divider — 36
      Example 16 – 3-to-8 Decoder — 40
   1.5 State Machines — 41
      Example 17 – A Moore Machine Sequence Detector — 42
      Example 18 – A Mealy Machine Sequence Detector — 45
      Example 19 – Door Lock Code — 48
   1.6 VHDL Package — 53
      Example 20 – Door Lock Code – Package — 54
   1.7 Multiple-Process VHDL Programs — 56
      Example 21 – The 7-Segment Display Module, *x7segb* — 56
   1.8 VHDL *while* Statement — 59
      Example 22 – GCD Algorithm – Part 1 — 59

2. Datapaths and Control Units — 64
   Example 23 – GCD Algorithm – Part 2 — 66
   Example 24 – Square Root Algorithm – Part 1 — 77

3. Integrating the Datapath and Control Unit — 89
   Example 25 – GCD Algorithm – Part 3 — 91
   Example 26 – Square Root Algorithm – Part 2 — 95

4. Memory — 99
- Example 27 – A VHDL ROM — 100
- Example 28 – Distributed RAM/ROM — 104
- Example 29 – A Stack — 109
- Example 30 – Block RAM — 116
- Example 31 – External RAM — 119
- Example 32 – External Flash Memory — 126

5. UART — 130
- Example 33 – Transmit Module — 132
- Example 34 – Receive Module — 139

6. VGA Controller — 146
- Example 35 – VGA-Stripes — 151
- Example 36 – VGA-PROM — 157
- Example 37 – Sprites in Block ROM — 163
- Example 38 – Screen Saver — 169
- Example 39 – External Video RAM — 175
- Example 40 – External Video Flash — 182

7. PS/2 Port — 188
- Example 41 – Keyboard — 191
- Example 42 – Mouse — 195

8. Graphics — 215
- Example 43 – Clearing the Screen — 216
- Example 44 – Plotting a Dot — 222
- Example 45 – Plotting a Line — 228
- Example 46 – Plotting a Star — 242
- Example 47 – Plotting a Circle — 250

9. Forth Core for FPGAs — 264
- Example 48 – FC16 Forth Core — 273
- Example 49 – Data Stack — 280
- Example 50 – Function Unit — 283
- Example 51 – Return Stack — 289
- Example 52 – FC16 Controller — 291
- Example 53 – GCD Forth Program — 301
- Example 54 – Square Root Forth Program — 308

Appendix A – Aldec Active-HDL Tutorial – Part 1 — 313
Appendix B – Test Bench and XPower Tutorial — 332
Appendix C – Making a Turnkey System — 342
Appendix D – VHDL Quick Reference Guide — 347

1

VHDL Building Blocks

1.1 Modeling Digital Circuits Using VHDL

The Nexys-2 board from Digilent that is shown in Fig. 1.1 contains the Xilinx Spartan3E-500 FG320 FPGA. This *field programmable gate array* consists of an array of 1,164 *configurable logic blocks* (CLBs) arranged as 46 rows by 34 columns. Each CLB contains four slices, each of which contains two 16 x 1 RAM look-up tables (LUTs), which can implement any combinational logic function of four variables. In addition to two look-up tables, each slice contains two D flip-flops which act as storage devices for bits. The CLB array is surrounded by an array of I/O blocks providing a maximum of 232 user I/O. The basic architecture of a Spartan-3E FPGA is shown in Fig. 1.2. This Spartan3E-500 FPGA also contains 368,640 bits of block RAM, twenty 18 x 18 multipliers, as well as four Digital Clock Manager (DCM) blocks. These DCMs are used to eliminate clock distribution delay and can also increase or decrease the frequency of the clock.

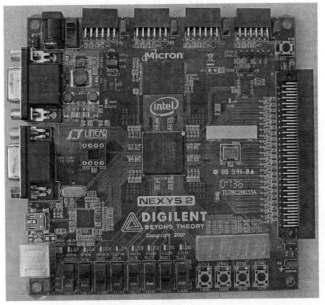

Figure 1.1 The Nexys-2 prototyping board

The traditional way of designing digital circuits is to draw logic diagrams containing gates and logic functions. However, the Spartan3E-500 FPGA used on the Nexys-2 board contains over 500,000 equivalent system gates. We say equivalent gates because as we have seen FPGAs really don't have AND and OR gates, but rather just RAM look-up tables. (Each slice does include two AND gates and two XOR gates as part of carry and arithmetic logic used when implementing arithmetic functions including addition and multiplication.). How can you draw schematic diagrams containing hundreds of thousands of gates? As programmable logic devices replaced TTL chips in new designs a new approach to digital design became necessary. Computer-aided tools are essential to designing digital circuits today. What has become clear over the last decade is that today's digital engineer designs digital systems by writing software! This is a major paradigm shift from the traditional method of designing digital systems.

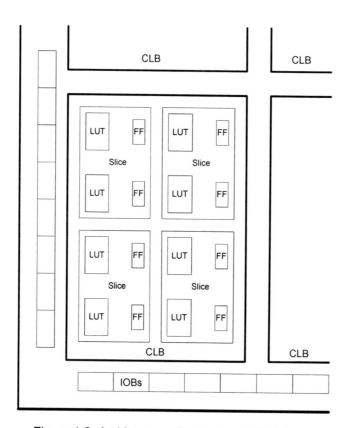

Figure 1.2 Architecture of a Spartan-3E FPGA

Today digital designers use *hardware description languages* (HDLs) to design digital systems. The most widely used HDLs are VHDL and Verilog. Both of these hardware description languages allow the user to design digital systems by writing a program that describes the behavior of the digital circuit. The program can then be used to both *simulate* the operation of the circuit and *synthesize* an actual implementation of the circuit in an FPGA or an application specific integrated circuit (ASIC).

VHDL is a double acronym: the V stands for VHSIC (Very High Speed Integrated Circuit) and the HDL stands for Hardware Description Language. The VHSIC program was launched in 1980 with the goal of achieving significant productivity gains in VLSI technology. During the 1980s VHDL was developed under government contract and became an IEEE standard (IEEE 1076) in 1987 and updated in 1993.

VHDL is based on the Ada programming language but it is *not* Ada. VHDL is a *hardware description language* that is designed to model digital logic circuits. It simply has the same syntax as the Ada programming language but the way it behaves is different. The prerequisite knowledge of VHDL assumed for this book can be obtained from our book *Learning By Example Using VHDL – Basic Digital Design Using a BASYS FPGA board* (L-BASYS), which is available from www.lbebooks.com. That book contains over sixty VHDL examples covering basic digital design of both combinational and sequential circuits. All of the examples in that book were simulated using the student version of the Aldec Active-HDL simulator, which is available free from http://www.aldec.com/Downloads/default.aspx. We will use this same simulator in this book. All of the examples in the L-BASYS book that run on the Digilent BASYS board will run on the Nexys-2 board by using the file *nexys2.ucf* (available from www.lbebooks.com) instead of *basys.ucf* to define the pin numbers.

To synthesize and implement your design to a Xilinx FPGA you must download the free Xilinx ISE WebPACK from www.xilinx.com. We will synthesize our VHDL designs to the Xilinx FPGA on the Digilent Nexys-2 prototyping board shown in Fig. 1.1. This board is available for under $100.00 from www.digilentinc.com. The board includes a Xilinx Spartan3E-500 FG320 FPGA, 8 slide switches, 4 pushbutton switches, 8 LEDs, and four 7-segment displays. It also has an on-board 50 MHz clock component, a serial port, a VGA port and a PS/2 port, onboard 16 Mbytes of RAM, and onboard 16 Mbytes of flash memory. There are connectors that allow the board to be interfaced to external circuits. After implementing your design using the WebPACK software the FPGA is programmed through a USB port using free Adept Suite Software that is downloaded from www.digilentinc.com.

In the L-BASYS book we described over sixty VHDL examples starting with simple logic equations and proceeding through examples of combinational and sequential logic and concluding with a discussion of state machines. In the rest of this chapter we will summarize these results by describing the important VHDL statements: *if* statements, *case* statements, and *for* loops. VHDL models described by these statements can not only be simulated but they can also be synthesized to hardware that is implemented in an FPGA. We will show that some VHDL statements such as the *while* loop can be simulated, but can *not* be synthesized. This will lead us to a discussion in Chapters 2 and 3 of how algorithms that are described by a *while* loop can be implemented in an FPGA using VHDL. These general methods of implementing algorithms in VHDL will be used throughout the rest of the book to illustrate the use of the main features of the Nexys-2 board including the VGA port, the serial port, the PS/2 port, the onboard 16 Mbytes of RAM, and the onboard 16 Mbytes of flash memory.

1.2 VHDL *if* Statement

The VHDL *if* statement is useful for describing the behavior of some digital systems. The general form of the *if* statement is

>[label:]   -- optional label
>**if** <boolean expression> **then**
>   <sequential statement>
>**elsif** <boolean expression> **then**
>   <sequential statement>
>**else**
>   <sequential statement>
>**end if** [label] ;   -- optional label

The *if* statement is an example of a sequential statement that must occur within a ***process***. The general form of the *process* statement is

>[label:]   -- optional label
>**process** (<sensitivity list>)
>   -- declarations
>   <variable declarations>
>**begin**
>   <sequential statement>
>**end process** [label];

Other sequential statements include the *case* statement and *for* loops that will be described in Sections 1.3 and 1.5. Sequential statements must be contained within a *process* and are executed in the order that they appear in the code.

The *process* block begins with an optional label followed by

```
process(<sensitivity_list>)
```

where the *sensitivity list* contains a list of all signals that will affect the outputs generated by the *process* block. You can think of the statements within the process as being "executed" any time one of the signals given in the sensitivity list changes.

An important point to remember is that signal assignment statements using the signal assignment operator <= within a process are not evaluated until the end of the process. Many times, however, we want the output of an assignment statement to be evaluated when it is executed. To do this we must define a *variable* within the process declarations area and use the variable assignment operator := instead of the signal assignment operator <=. At the end of the process the value of the variable is typically assigned to some signal or output using the signal assignment operator.

There can be more than one process within an architecture of a VHDL program. While the statements within a process execute sequentially, multiple processes within an architecture execute concurrently.

Example 1: 2-to-1 Multiplexer

As a first example of using VHDL consider the *N*-line 2 x 1 multiplexer shown in Fig. 1.3. A VHDL program for this multiplexer is given in Listing 1.1. The first line is a VHDL comment line that begins with --. The next two lines must be at the start of all VHDL programs. It indicates that all of the contents of the library file *std_logic_1164.vhd* located in the IEEE directory are to be included in this program. This system file contains definitions of the basic VHDL data types and logic functions.

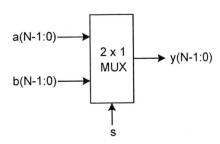

Figure 1.3 An *N*-line 2-to-1 multiplexer

Listing 1.1 mux2g.vhd

```vhdl
-- Example 1a: Generic 2-to-1 MUX using a parameter
library IEEE;
use IEEE.STD_LOGIC_1164.all;

entity mux2g is
    generic (N:integer);
    port (
        a: in STD_LOGIC_VECTOR(N-1 downto 0);
        b: in STD_LOGIC_VECTOR(N-1 downto 0);
        s: in STD_LOGIC;
        y: out STD_LOGIC_VECTOR(N-1 downto 0)
    );
end mux2g;

architecture mux2g of mux2g is
begin
    process(a, b, s)
      begin
        if s = '0' then
            y <= a;
        else
            y <= b;
        end if;
      end process;
end mux2g;
```

All VHDL programs are made up of two basic components: an *entity* and an *architecture*. The entity in Listing 1.1 is named *mux2g* and describes the input and output signals shown in Fig. 1.3. The optional *generic* statement in the second line of the entity defines *N* to be the size of the data busses *a*, *b*, and *y* in Fig. 1.3. In Listing 1.3 *a*, *b*, and *y* are defined to be of type STD_LOGIC_VECTOR and *s* is defined to be of type STD_LOGIC. The type STD_LOGIC can have any of the nine values shown in Table 1.1.

Table 1.1 Values of type STD_LOGIC

Value	Meaning
'U'	Uninitialized
'X'	Forcing Unknown
'0'	Forcing 0
'1'	Forcing 1
'Z'	High Impedance
'W'	Weak Unknown
'L'	Weak 0
'H'	Weak 1
'-'	Don't care

The architecture of a VHDL program describes what the component defined by the entity does. In Listing 1.1 the behavior of the multiplexer is described by the *if...then...else* statement. Note that <= is used in VHDL as the signal assignment operator. The *if...then...else* statement is an example of a VHDL sequential statement. All sequential statements must occur within a *process*. The process in Listing 1.1 is defined by the statement *process (a, b, s)*. The signals *a*, *b*, and *s* within the parentheses of the *process* statement are called the *sensitivity list*. The statements within a process are executed any time that any values of the signals in the sensitivity list change. All signals whose values are assigned within a process (e.g. *y* in Listing 1.1) are updated at the end of a process. An architecture can have any number of processes. These processes execute concurrently.

In Listing 1.1 note the use of the *generic* statement that defines the bus width *N*. This value is set when the multiplexer is instantiated by including a *generic map* statement as in the following example of an 8-line 2-to-1 multiplexer called *M8*.

```
M8 : mux2g
     generic map(
         N => 8
     )
     port map(
         a => a,
         b => b,
         s => s,
         y => y
     );
```

We will always use upper-case names for parameters.

A complete VHDL program for this 8-line 2-to-1 multiplexer is given in Listing 1.2 and its simulation is shown in Fig. 1.4. The tutorial in Appendix A will show how to generate this simulation using Aldec Active-HDL.

Listing 1.2 mux28.vhd

```vhdl
-- Example 1b: 8-line 2-to-1 MUX using a parameter
library IEEE;
use IEEE.STD_LOGIC_1164.all;

entity mux28 is
    port(
        s : in STD_LOGIC;
        a : in STD_LOGIC_VECTOR(7 downto 0);
        b : in STD_LOGIC_VECTOR(7 downto 0);
        y : out STD_LOGIC_VECTOR(7 downto 0)
    );
end mux28;

architecture mux28 of mux28 is
    component mux2g
    generic(
        N : integer);
    port(
        a : in std_logic_vector((N-1) downto 0);
        b : in std_logic_vector((N-1) downto 0);
        s : in std_logic;
        y : out std_logic_vector((N-1) downto 0));
    end component;

begin
M8 : mux2g
    generic map(
        N => 8
    )
    port map(
        a => a,
        b => b,
        s => s,
        y => y
    );
end mux28;
```

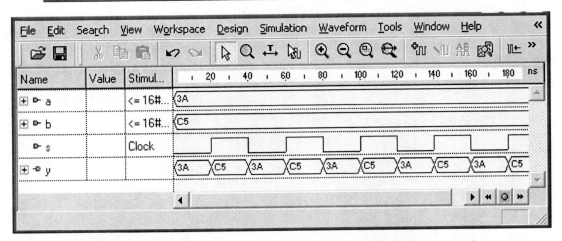

Figure 1.4   Simulation result from the VHDL program in Listing 1.2

Example 2: Registers

A logic diagram for an *N*-bit register is shown in Fig. 1.5 and the VHDL program for this register is shown in Listing 1.3. The behavior of this register is that if *clr* = '1' then all elements of *q(N-1:0)* become zero asynchronously. Otherwise, if *load* = '1' then on the rising edge of the clock signal *clk* the input *d(N-1:0)* gets latched into the output *q(N-1:0)*.

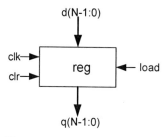

Figure 1.5 An *N*-bit register

The way to tell VHDL that you want registered outputs of signals is to describe the behavior in a process that includes an *if* statement containing the phrase *clk'event* **and** *clock* = '1' or *rising_edge(clk)*. While the second phrase is more descriptive the phase *clk'event* **and** *clock* = '1' is more common usage in VHDL programs. It means that if there was an event on the clock signal (the clock changes value) and after the event the clock is high (i.e. it was a rising edge) then the *if* phrase will be true. The process sensitivity list in Listing 1.3 contains the two inputs *clk* and *clr*. It is only when either of these two inputs changes that we want the process to execute. Note that if *clr* is equal to '1' then all elements of *q(N-1:0)* will become zero immediately (asynchronously). This can be accomplished using the statement

```
q <= (others => '0');
```

If *clr* is not '1' then on the rising edge of the clock signal, *clk*, if the value of *load* is '1', the value of *q(N-1:0)* will be set to the current value of *d(N-1:0)*, which is exactly the behavior of the *N*-bit register. A simulation of Listing 1.3 is shown in Fig. 1.6.

Note the use of the **generic** statement in Listing 6.3, which defines the bus width *N* to have a default value of 8. This value can be overridden when the register is instantiated by including *generic map* clause in the *port map* statement as in the following example of a 16-bit register called *aReg*.

```
aReg: reg
    generic map(
        N => 16
    )
    port map(
        load => load,
        d => d,
        clk => clk,
        clr => clr,
        q => q
    );
```

VHDL *if* Statement

Listing 1.3 reg.vhd

```vhdl
-- Example 2: N-bit register with clear and load
library IEEE;
use IEEE.STD_LOGIC_1164.all;

entity reg is
      generic(N : integer := 8);
      port(
            load : in STD_LOGIC;
            d : in STD_LOGIC_VECTOR(N-1 downto 0);
            clk : in STD_LOGIC;
            clr : in STD_LOGIC;
            q : out STD_LOGIC_VECTOR(N-1 downto 0)
         );
end reg;

architecture reg of reg is
begin
      -- N-bit register with load
      process(clk, clr)
      begin
            if clr = '1' then
                  q <= (others => '0');
            elsif clk'event and clk = '1' then
                  if load = '1' then
                        q <= d;
                  end if;
            end if;
      end process;
end reg;
```

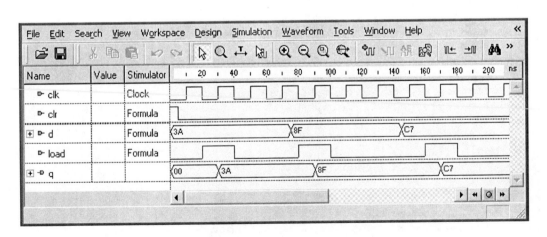

Figure 1.6 Simulation of the VHDL program in Listing 1.3

Example 3: Debounce Pushbuttons

When you press any of the pushbuttons on the Nexys-2 board they may bounce slightly for a few milliseconds before settling down. This means that instead of the input to the FPGA going from 0 to 1 cleanly it may bounce back and forth between 0 and 1 for a few milliseconds. This can be a serious problem in sequential circuits where actions take place on the rising edge of a clock signal. Because this clock signal changes much faster than the switch bouncing it is possible for erroneous values to be latched into registers. For this reason it is necessary to debounce the pushbutton switches when you use them in sequential circuits.

The circuit shown in Fig. 1.7 that uses three D flip-flops can be used to debounce a pushbutton input signal, *inp*. The frequency of the input clock, *cclk*, must be low enough that the switch bouncing is over before three clock periods. We will normally use a 190 Hz frequency for *cclk*. Listing 1.4 is a VHDL program that implements four versions of the debounce circuit in Fig. 1.7 – one for each of the four pushbuttons on the Nexys-2 board.

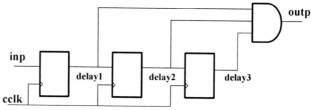

Figure 1.7 Debounce circuit

Listing 1.4 debounce4.vhd

```vhdl
-- Example 3: Debounce pushbuttons
library IEEE;
use IEEE.std_logic_1164.all;

entity debounce4 is
    port (
        cclk, clr: in std_logic;
        inp: in std_logic_vector(3 downto 0);
        outp: out std_logic_vector(3 downto 0)
    );
end debounce4;

architecture debounce4 of debounce4 is
signal delay1, delay2, delay3: std_logic_vector(3 downto 0);
begin
    process(cclk, clr)
    begin
        if clr = '1' then
                delay1 <= "0000";
                delay2 <= "0000";
                delay3 <= "0000";
        elsif cclk'event and cclk='1' then
                delay1 <= inp;
                delay2 <= delay1;
                delay3 <= delay2;
        end if;
    end process;

    outp <= delay1 and delay2 and delay3;
end debounce4;
```

You can understand how this debounce circuit works by looking at the simulation shown in Fig. 1.8 where a bouncing input signal is used as a stimulator for *inp*(0). We have indicated bouncing on both pressing and releasing the button. Note that the resulting ouput signal *outp*(0) is a clean signal with all bouncing effects removed. The output will not go high until the input has been high for three clock cycles. The ouput will stay low as long as any bounces do not remain high for three clock cycles. Thus, it is important to use a low frequency for *cclk* to make sure all debounces are eliminated.

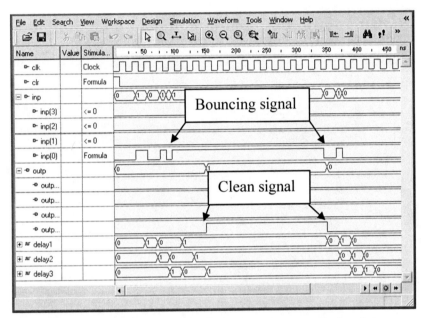

Figure 1.8 Simulation of the VHDL program in Listing 1.4

Example 4: Clock Pulse

A very useful circuit that will produce a single clean clock pulse is shown in Fig. 1.9. The only difference from the debounce circuit in Fig. 1.7 is that the complement of *delay3* is the last input to the AND gate. Listing 1.5 is a VHDL program for this clock pulse circuit and its simulation is shown in Fig. 1.10.

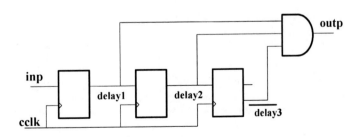

Figure 1.9 Clock pulse circuit

Listing 1.5 clock_pulse.vhd

```vhdl
-- Example 4: clock pulse
library IEEE;
use IEEE.std_logic_1164.all;

entity clock_pulse is
      port (
              inp, cclk, clr: in std_logic;
              outp: out std_logic
              );
end clock_pulse;

architecture clock_pulse_arch of clock_pulse is
signal delay1, delay2, delay3: std_logic;
begin
     process(cclk, clr)
     begin
        if clr = '1' then
              delay1 <= '0';
              delay2 <= '0';
              delay3 <= '0';
        elsif cclk'event and cclk='1' then
              delay1 <= inp;
              delay2 <= delay1;
              delay3 <= delay2;
        end if;
     end process;
     outp <= delay1 and delay2 and (not delay3);
end clock_pulse_arch;
```

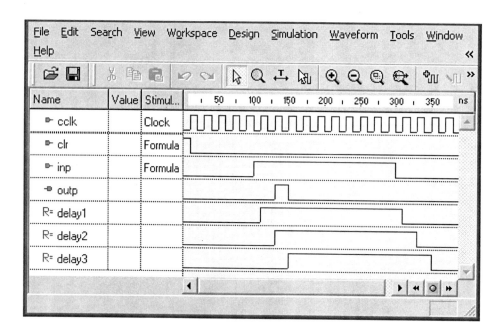

Figure 1.10 Simulation of the Verilog program in Listing 1.5

Example 5: Counters

We can make a general *N*-bit counter by using the *generic* statement as shown in Listing 1.6, which has a default value of 8. A simulation of an 8-bit counter that will count from 0 – 255 (or 00 – FF hex) is shown in Fig. 1.11. (A count from about 00 – B7 is shown in the simulation.)

Listing 1.6 counter.vhd

```vhdl
-- Example 5: N-bit counter
library IEEE;
use IEEE.STD_LOGIC_1164.all;
use IEEE.STD_LOGIC_unsigned.all;

entity counter is
        generic(N : integer := 8);
        port(
                clr : in STD_LOGIC;
                clk : in STD_LOGIC;
                q : out STD_LOGIC_VECTOR(N-1 downto 0)
             );
end counter;

architecture counter of counter is
signal count: STD_LOGIC_VECTOR(N-1 downto 0);
begin
        -- N-bit counter
        process(clk, clr)
        begin
                if clr = '1' then
                        count <= (others => '0');
                elsif clk'event and clk = '1' then
                        count <= count + 1;
                end if;
        end process;

        q <= count;
end counter;
```

. To instantiate a 16-bit counter from Listing 1.6 that would count from 0 – 65535 (or 0000 – FFFF hex) you would use something like

```vhdl
cnt16: counter
    generic map(
        N => 16
    )
    port map(
        clr => clr,
        clk => clk,
        q => q
    );
```

Note in Listing 1.6 that we have included the additional *use* statement

```
use IEEE.STD_LOGIC_unsigned.all;
```

This statement will include the library file *unsigned.vhd* in the project. This is required in order to use the + sign to implement the counter by adding 1 to the signal *count*.

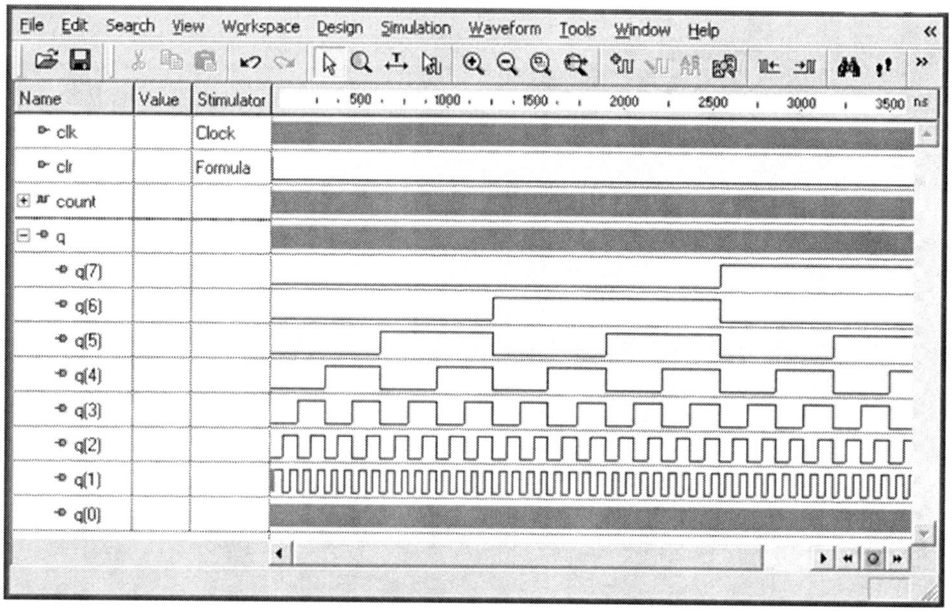

Figure 1.11 Simulation of the 8-bit counter in Listing 1.6

Example 6: Clock Divider

The simulation in Fig. 1.11 shows that the outputs $q(i)$ of a counter are square waves where the output $q(0)$ has a frequency half of the clock frequency, the output $q(1)$ has a frequency half of $q(0)$, etc. Thus, a counter can be used to divide the frequency *f* of a clock, where the frequency of the output $q(i)$ is $f_i = f/2^{i+1}$. The frequencies and periods of the outputs of a 24-bit counter driven by a 50 MHz clock are shown in Table 1.2. Note in Table 1.2 that the output $q(0)$ has a frequency of 25 MHz, the output $q(17)$ has a frequency of 190.73 Hz, and the output $q(23)$ has a frequency of 2.98 Hz.

The VHDL program shown in Listing 1.7 is a 24-bit counter that has three outputs, a 25 MHz clock (*clk25*), a 190 Hz clock (*clk190*), and a 3 Hz clock (*clk3*). You can modify this *clkdiv* component to produce any output frequency given in Table 1.2. We will use such a clock divider component in many of our top-level designs.

Table 1.2 Clock divide frequencies

q(i)	Frequency (Hz)	Period (ms)
i	50000000.00	0.00002
0	25000000.00	0.00004
1	12500000.00	0.00008
2	6250000.00	0.00016
3	3125000.00	0.00032
4	1562500.00	0.00064
5	781250.00	0.00128
6	390625.00	0.00256
7	195312.50	0.00512
8	97656.25	0.01024
9	48828.13	0.02048
10	24414.06	0.04096
11	12207.03	0.08192
12	6103.52	0.16384
13	3051.76	0.32768
14	1525.88	0.65536
15	762.94	1.31072
16	381.47	2.62144
17	190.73	5.24288
18	95.37	10.48576
19	47.68	20.97152
20	23.84	41.94304
21	11.92	83.88608
22	5.96	167.77216
23	2.98	335.54432

Listing 1.7 clkdiv.vhd

```vhdl
-- Example 6: Clock divider
library IEEE;
use IEEE.STD_LOGIC_1164.all;
use IEEE.STD_LOGIC_unsigned.all;

entity clkdiv is
        port(
             mclk : in STD_LOGIC;
             clr : in STD_LOGIC;
             clk25 : out STD_LOGIC;
             clk190 : out STD_LOGIC;
             clk3 : out STD_LOGIC
           );
end clkdiv;

architecture clkdiv of clkdiv is

signal q: STD_LOGIC_VECTOR(23 downto 0);
```

Listing 1.7 (cont.) clkdiv.vhd

```vhdl
begin
      -- clock divider
      process(mclk, clr)
      begin
            if clr = '1' then
                  q <= X"000000";
            elsif mclk'event and mclk = '1' then
                  q <= q + 1;
            end if;
      end process;

      clk25  <= q(0);           -- 25 MHz
      clk190 <= q(17);          -- 190 Hz
      clk3   <= q(23);          -- 3 Hz

end clkdiv;
```

Example 7: Comparators

The easiest way to implement a comparator in VHDL is to use the relational and logical operators shown in Table 1.3. An example of using these to implement a 4-bit comparator is shown in Listing 1.8. A simulation of this program is shown in Fig. 1.12.

Note in the process in Listing 1.8 we set the values of *gt*, *eq*, and *lt* to zero before the *if* statements. This is important to make sure that each output has a value assigned to it. If you don't do this then VHDL will assume you don't want the value to change and will include a latch in your system. Your circuit will then not be a combinational circuit.

Again in Listing 1.8 we must include the additional *use* statement

```
use IEEE.STD_LOGIC_unsigned.all;
```

in order to use the relational operators.

Table 1.3 Relational and Logical Operators

Operator	Meaning
=	Logical equality
/=	Logical inequality
<	Less than
<=	Less than or equal
>	Greater than
>=	Greater than or equal
not	Logical negation
and	Logical AND
or	Logical OR

Listing 1.8 comp4.vhd

```vhdl
-- Example 7: 4-bit comparator using relational operators
library IEEE;
use IEEE.STD_LOGIC_1164.all;
use IEEE.STD_LOGIC_unsigned.all;

entity comp4 is
    port(
            x : in STD_LOGIC_VECTOR(3 downto 0);
            y : in STD_LOGIC_VECTOR(3 downto 0);
            gt : out STD_LOGIC;
            eq : out STD_LOGIC;
            lt : out STD_LOGIC
        );
end comp4;

architecture comp4 of comp4 is
begin
    process(x,y)
    begin
        gt <= '0';
        eq <= '0';
        lt <= '0';
        if x > y then
            gt <= '1';
        end if;
        if x = y then
            eq <= '1';
        end if;
        if x < y then
            lt <= '1';
        end if;
    end process;
end comp4;
```

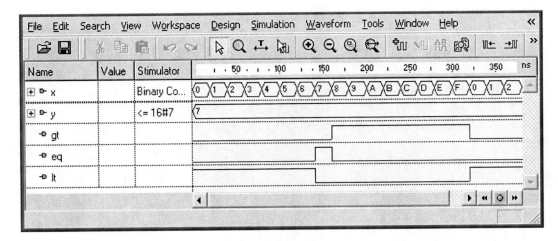

Figure 1.12 Simulation of the VHDL program in Listing 1.8

1.3 VHDL *case* Statement

The VHDL *case* statement is useful for describing the behavior of some digital systems. The general form of the *case* statement is

> [label:]   -- optional label
> **case** <expression> **is**
>   **when** <choices> => <sequential statement>
>   **when** <choices> => <sequential statement>
>   ------------
> **end case** [label] ;   -- optional label

The *case* statement is an example of a sequential statement that must occur within a **process**. In the *case* statement all <choices> of <expression> must be included. Inasmuch as <expression> is often of type STD_LOGIC that has 9 possible values, the last choice in a *case* statement should be

> **when others** => <sequential statement>

where *null* could be used for <sequential statement>.

Example 8: 4-to-1 Multiplexer

In this example we will show how to design a 4-to-1 multiplexer using the VHDL *case* statement. The VHDL program shown in Listing 1.9 implements the *N*-line 4-to-1 MUX shown in Fig. 1.13. The *case* statement in Listing 1.9 directly implements the definition of a 4-to-1 MUX. A typical line in the *case* statement, such as

> **when** "10"   => z <= c;

will assign the value of *c* to the output *y* when the input value $s(1:0)$ is equal to "10".

In the *case* statement the alternative value following the *when* in each line represents the value of the *case* parameter, in this case the 2-bit input *sel*. Note that STD_LOGIC_VECTOR constants are enclosed in double quotes (").

All *case* statements should include a default line *when others* as shown in Listing 1.9. This is because all cases need to be covered and while it looks as if we covered all cases in Listing 1.9, recall that VHDL actually defines *nine* possible values for each element of a STD_LOGIC_VECTOR (see Table 1.1). The simulation of the program in Listing 1.9 is shown in Fig. 1.14.

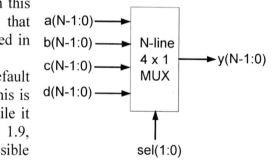

Figure 1.13 An *N*-line 4 x 1 multiplexer

VHDL case Statement

Listing 1.9 mux4g.vhd

```vhdl
-- Example 8: 4-to-1 multiplexer
library IEEE;
use IEEE.std_logic_1164.all;

entity mux4g is
    generic(N : integer := 4);
    port (
        a: in STD_LOGIC_VECTOR (N-1 downto 0);
        b: in STD_LOGIC_VECTOR (N-1 downto 0);
        c: in STD_LOGIC_VECTOR (N-1 downto 0);
        d: in STD_LOGIC_VECTOR (N-1 downto 0);
        sel: in STD_LOGIC_VECTOR (1 downto 0);
        y: out STD_LOGIC_VECTOR (N-1 downto 0)
    );
end mux4g;

architecture mux4g_arch of mux4g is
begin
  process (sel, a, b, c, d)
  begin
    case sel is
      when "00"   => y <= a;
      when "01"   => y <= b;
      when "10"   => y <= c;
      when others => y <= d;
    end case;
  end process;
end mux4g_arch;
```

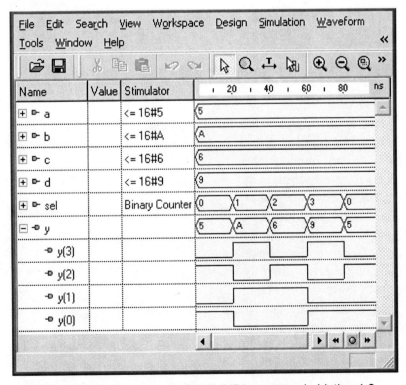

Figure 1.14 Simulation of the VHDL program in Listing 1.9

Example 9: 7-Segment Decoder

The Nexys-2 board has four common-anode 7-segment displays. This means that all the anodes are tied together and connected through a *pnp* transistor to +3.3V. A different FPGA output pin is connected through a 100 Ω current-limiting resistor to each of the cathodes, $a - g$, plus the decimal point. In the common-anode case, an output 0 will turn on a segment and an output 1 will turn it off. The table shown in Fig. 1.15 shows output cathode values for each segment $a - g$ needed to display all hex values from $0 - F$.

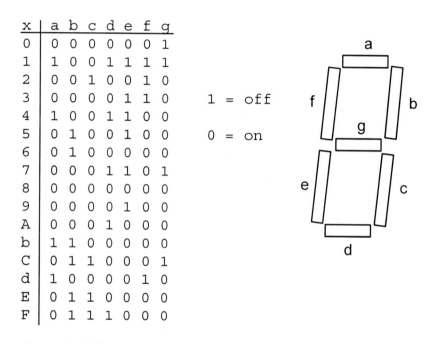

x	a	b	c	d	e	f	g
0	0	0	0	0	0	0	1
1	1	0	0	1	1	1	1
2	0	0	1	0	0	1	0
3	0	0	0	0	1	1	0
4	1	0	0	1	1	0	0
5	0	1	0	0	1	0	0
6	0	1	0	0	0	0	0
7	0	0	0	1	1	0	1
8	0	0	0	0	0	0	0
9	0	0	0	0	1	0	0
A	0	0	0	1	0	0	0
b	1	1	0	0	0	0	0
C	0	1	1	0	0	0	1
d	1	0	0	0	0	1	0
E	0	1	1	0	0	0	0
F	0	1	1	1	0	0	0

1 = off
0 = on

Figure 1.15 Segment values required to display hex digits 0 – F

The problem is to design a *hex to 7-segment decoder*, called *hex7seg*, that is shown in Fig. 1.16. The input is a 4-bit hex number, $x(3:0)$, and the outputs are the 7-segment values $a - g$ given by the truth table in Fig. 1.15.

The VHDL program shown in Listing 1.10 is a hex-to-seven-segment decoder that converts a 4-bit input hex digit, $0 - F$, to the appropriate 7-segment codes, $a - g$. The *case* statement in Listing 1.10 directly implements the truth table in Fig. 1.15. In the array a_to_g the value a_to_g(6) corresponds to segment a and the value a_to_g(0) corresponds to segment g.

Figure 1.16 A hex to 7-segment decoder

A simulation of Listing 1.10 is shown in Fig. 1.17. Note that the simulation agrees with the truth table in Fig. 1.15.

Listing 1.10 hex7seg.vhd

```vhdl
-- Example 9a: Hex to 7-segment decoder: a - g active low
library IEEE;
use IEEE.STD_LOGIC_1164.all;

entity hex7seg is
        port(
                x : in STD_LOGIC_VECTOR(3 downto 0);
                a_to_g : out STD_LOGIC_VECTOR(6 downto 0)
            );
end hex7seg;

architecture hex7seg of hex7seg is
begin
    process(x)
    begin
        case x is
            when "0000" => a_to_g <= "0000001";   --0
            when "0001" => a_to_g <= "1001111";   --1
            when "0010" => a_to_g <= "0010010";   --2
            when "0011" => a_to_g <= "0000110";   --3
            when "0100" => a_to_g <= "1001100";   --4
            when "0101" => a_to_g <= "0100100";   --5
            when "0110" => a_to_g <= "0100000";   --6
            when "0111" => a_to_g <= "0001101";   --7
            when "1000" => a_to_g <= "0000000";   --8
            when "1001" => a_to_g <= "0000100";   --9
            when "1010" => a_to_g <= "0001000";   --A
            when "1011" => a_to_g <= "1100000";   --b
            when "1100" => a_to_g <= "0110001";   --C
            when "1101" => a_to_g <= "1000010";   --d
            when "1110" => a_to_g <= "0110000";   --E
            when others => a_to_g <= "0111000";   --F
        end case;
    end process;
end hex7seg;
```

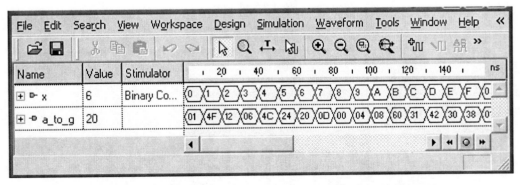

Figure 1.17 Simulation of the Verilog program in Listing 1.10

Selected Signal Assignment Statement

It is possible to design the 7-segment decoder shown in Fig. 1.16 using a *selected signal assignment statement* as shown in Listing 1.11. Note that this *with...select...when* statement is a concurrent statement that does not need to be in a process. Listing 1.11 will produce the same simulation shown in Fig. 1.17.

Listing 1.11 seg7dec.vhd

```vhdl
-- Example 9b: Hex to 7-seg decoder: with..select..when
library IEEE;
use IEEE.std_logic_1164.all;

entity seg7dec is
    port (x: in STD_LOGIC_VECTOR(3 downto 0);
          a_to_g: out STD_LOGIC_VECTOR(6 downto 0));
end seg7dec;

architecture seg7dec of seg7dec is
begin
-- seg7dec
    with x select
        a_to_g <= "1001111" when "0001",   --1
                  "0010010" when "0010",   --2
                  "0000110" when "0011",   --3
                  "1001100" when "0100",   --4
                  "0100100" when "0101",   --5
                  "0100000" when "0110",   --6
                  "0001111" when "0111",   --7
                  "0000000" when "1000",   --8
                  "0000100" when "1001",   --9
                  "0001000" when "1010",   --A
                  "1100000" when "1011",   --b
                  "0110001" when "1100",   --C
                  "1000010" when "1101",   --d
                  "0110000" when "1110",   --E
                  "0111000" when "1111",   --F
                  "0000001" when others;   --0
end seg7dec;
```

Example 10: Arithmetic Logic Unit (ALU)

In this example we will show how to design an arithmetic logic unit (ALU) using VHDL. Similar to the select lines on a multiplexer that select which of the many inputs will be selected as output, the ALU will also have a set of select lines that will control which operation will be used. Table 1.4 lists some common arithmetic and logic operations that we will implement in an ALU.

This ALU will output *a* directly when the select line, *alusel* is "000" and perform addition when the select line, *alusel* is "001". An *alusel* of "010" will perform the subtraction a – b, and an *alusel* of "011" will perform the subtraction b – a. Logical operations NOT, AND, OR, and XOR are output for *alusel* of "100" through "111".

Fig. 1.18 shows the logic symbol for a 4-bit ALU. Since there are eight operations, *alusel* is 3 bits. In addition to the 4-bit output $y(3:0)$, four flags are also output from the ALU. The carry flag, *cf*, and the overflow flag, *ovf*, are the carry/borrow and overflow flags from the addition and subtraction operations. The zero flag, *zf*, is 1 if the output, *y*, is "0000". The negative flag, *nf*, is 1 if the sign bit, $y(3)$ is 1. Listing 1.12 shows the VHDL program for this ALU and Fig. 1.19 shows a simulation of this program.

Table 1.4 ALU operations

alusel(2:0)	Function	Output
0 0 0	Pass a	a
0 0 1	Addition	a + b
0 1 0	Subtraction-1	a - b
0 1 1	Subtraction-2	b - a
1 0 0	Invert	NOT a
1 0 1	And	a AND b
1 1 0	Or	a OR b
1 1 1	Xor	a XOR b

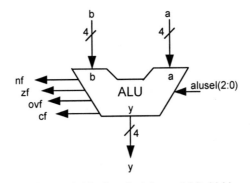

Figure 1.18 Symbol for a 4-bit ALU

Figure 1.19 Simulation of the VHDL program in Listing 1.12

Listing 1.12 alu.vhd

```vhdl
-- Example 10: alu
library IEEE;
use IEEE.STD_LOGIC_1164.all;
use IEEE.STD_LOGIC_unsigned.all;

entity alu is
       port(
              alusel : in STD_LOGIC_VECTOR(2 downto 0);
              a : in STD_LOGIC_VECTOR(3 downto 0);
              b : in STD_LOGIC_VECTOR(3 downto 0);
              nf : out STD_LOGIC;
              zf : out STD_LOGIC;
              cf : out STD_LOGIC;
              ovf : out STD_LOGIC;
              y : out STD_LOGIC_VECTOR(3 downto 0)
           );
end alu;

architecture alu of alu is
begin
       process(a,b,alusel)
       variable temp: STD_LOGIC_VECTOR(4 downto 0);
       variable yv: STD_LOGIC_VECTOR(3 downto 0);
       variable cfv, zfv: STD_LOGIC;
       begin
              cf <= '0';
              ovf <= '0';
              temp := "00000";
              zfv := '0';
              case alusel is
                     when "000" =>                                    -- pass
                            yv := a;
                     when "001" =>                                    -- a + b
                            temp := ('0' & a) + ('0' & b);
                            yv := temp(3 downto 0);
                            cfv := temp(4);
                            ovf <= yv(3) xor a(3) xor b(3) xor cfv;
                            cf <= cfv;
                     when "010" =>                                    -- a - b
                            temp := ('0' & a) -    ('0' & b);
                            yv := temp(3 downto 0);
                            cfv := temp(4);
                            ovf <= yv(3) xor a(3) xor b(3) xor cfv;
                            cf <= cfv;
                     when "011" =>                                    -- b - a
                            temp := ('0' & b) -    ('0' & a);
                            yv := temp(3 downto 0);
                            cfv := temp(4);
                            ovf <= yv(3) xor a(3) xor b(3) xor cfv;
                            cf <= cfv;
                     when "100" =>                                    -- NOT
                            yv := not a;
                     when "101" =>                                    -- AND
                            yv := a and b;
                     when "110" =>                                    -- OR
                            yv := a or b;
```

Listing 1.12 (cont.) alu.vhd
```
                when "111" =>                    -- XOR
                    yv := a xor b;
                when others =>
                    yv := a;
            end case;
            for i in 0 to 3 loop
                zfv := zfv or yv(i);  -- zfv = '0' if all yv(i) = '0'
            end loop;
            y <= yv;
            zf <= not zfv;
            nf <= yv(3);
    end process;
end alu;
```

1.4 VHDL *for* Loop

The VHDL *for* loop is useful for describing the behavior of some digital systems. The general form of the *for* loop is

> [label:]   -- optional label
> **for** <identifier> **in** <range> **loop**
>   <sequential statement>
> **end loop** [label] ;   -- optional label

The *for* loop is an example of a sequential statement that must occur within a **process**. When the *for* loop is executed new hardware is generated each time through the loop. In this sense you can think of the *for* loop as "unwrapping" to produce a kind of cascading hardware. This will become apparent in Examples 11 – 18 in this section.

Example 11: 4-Input Gates

In this example we will show how to use the VHDL *for* loop to describe the behavior of 4-input gates. Consider an AND gate with four inputs as shown in Fig. 1.20. We can write the logic equation for this gate as

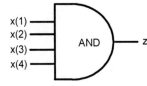

Figure 1.20
4-input AND gate.

```
z <= x(1) and x(2) and x(3) and x(4);
```

It is clear that we can implement this 4-input AND gate using three 2-input AND gates as shown in Figure 1.21. We can use the VHDL *for loop* to describe this behavior by writing

```
zv := x(1);
for i in 2 to 4 loop
    zv := zv and x(i);
end loop;
z <= zv;
```

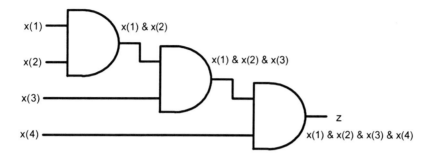

Figure 1.21 Implementing a 4-input AND gate using three 2-input AND gates

Note how the VHDL *for* loop unwraps the statement `zv := zv and x(i)` to produce the circuit shown in Fig. 1.21. A VHDL program that uses this *for* loop is shown in Listing 1.13 and the resulting simulation is shown in Fig. 1.22.

Listing 1.13: and4.vhd

```
-- Example 30: 4-input AND gate - VHDL for loop
library IEEE;
use IEEE.STD_LOGIC_1164.all;

entity and4 is
        port(
                x : in STD_LOGIC_VECTOR(4 downto 1);
                z : out STD_LOGIC
            );
end and4;

architecture and4 of and4 is
begin
and4_1: process(x)
variable zv: STD_LOGIC;
begin
        zv := x(1);
        for i in 2 to 4 loop
                zv := zv and x(i);
        end loop;
        z <= zv;
end process and4_1;
end and4;
```

The *for* loop is an example of a *procedural*, or *sequential*, statement. Procedural statements must be contained within a *process* and are executed in the order that they appear in the code. In Listing 5.1 the sensitivity list contains the array $x(4:1)$ so that a change in any of the four inputs will affect the output z.

Recall that signal assignment statements using the signal assignment operator <= within a process are not evaluated until the end of the process. Many times, however, we want the output of an assignment statement to be evaluated when it is executed. This is the case within the *for* loop in Listing 1.13 where *zv* must be evaluated and remembered each time through the *for* loop. To do this *zv* must be defined as a STD_LOGIC *variable* within the process and the variable assignment operator := must be used. At the end of the *for* loop the output *z* is assigned the value of the variable *zv* using the signal assignment operator.

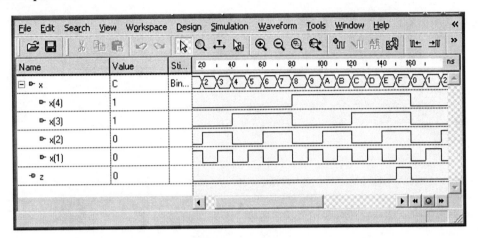

Figure 1.22 Simulation of VHDL code in Listing 1.13

Example 12: Binary-to-BCD Converter

In this example we will show how to design an 8-bit and a 14-bit binary-to-BCD converter using VHDL *for* loops. For an 8-bit binary-to-BCD converter a 2-digit hex number, 00 – FF, will be converted to the corresponding 3-digit BDC number, 000 – 255. This will be useful if you want to display a hex result in decimal. One way to do this is to use the so-called *shift and add 3 algorithm*.

Shift and Add 3 Algorithm

To see how the *shift and add 3 algorithm* works consider the example shown in Fig. 1.23. The 2-digit hex number to be converted is written as an 8-bit binary number in the two right-most columns in Fig. 5.4. In this case the hex number to be converted is FF so that 11111111 is written in the *Start* row. The next three columns from the right will end up containing the converted three BCD digits. They are labeled *Hundreds*, *Tens*, and *Units*, and will end up containing the three BCD digits 255 in this case.

The *shift and add 3 algorithm* consists of the following steps.

1. Shift the binary number left one bit.
2. If 8 shifts have taken place, the BCD number is in the *Hundreds*, *Tens*, and *Units* column.
3. If the binary value in any of the BCD columns is 5 or greater, add 3 to that value in that BCD column.
4. Go to 1.

Follow this algorithm in Fig. 1.23 to see how the hex value FF gets converted to the BCD value 255.

How can we design a logic circuit to perform the binary-to-BCD algorithm described in Fig. 1.23? The easiest way is to write behavioral VHDL statements that directly implements the *shift and add 3* algorithm. In Fig. 1.23 we have defined the input array $b(7:0)$, the output array $p(9:0)$, and a composite array $z(17:0)$ that covers both p and b.

Operation	Hundreds	Tens	Units	Binary
b				7 4 3 0
HEX				F F
Start				1 1 1 1 1 1 1 1
Shift 1			1	1 1 1 1 1 1 1
Shift 2			1 1	1 1 1 1 1 1
Shift 3			1 1 1	1 1 1 1 1
Add 3			1 0 1 0	1 1 1 1 1
Shift 4		1	0 1 0 1	1 1 1 1
Add 3		1	1 0 0 0	1 1 1 1
Shift 5		1 1	0 0 0 1	1 1 1
Shift 6		1 1 0	0 0 1 1	1 1
Add 3		1 0 0 1	0 0 1 1	1 1
Shift 7	1	0 0 1 0	0 1 1 1	1
Add 3	1	0 0 1 0	1 0 1 0	1
Shift 8	1 0	0 1 0 1	0 1 0 1	
BCD	2	5	5	
p	9 8	7 4	3 0		
z	17 16	15 12	11 8	7 4	3 0

Figure 1.23 Definition of the arrays **b**, **p**, and **z**

Using Fig. 1.23 as a guide we can implement the *shift and add 3* algorithm directly using the VHDL program shown in Listing 1.14. Note that we first clear the z array and then shift b three places to the left in z. We then go through a *for* loop 5 times, shifting z left 1 bit each time, to make a total of 8 shifts left. Each time through the *for* loop we check to see if the *units* or *tens* are 5 or greater, and if so, we add 3. When we exit the *for* loop the BCD output p will be in $z(17:8)$. A simulation of the program in Listing 1.14 is shown in Fig. 1.24.

The main *for* loop in Listing 1.14 unwraps to produce the hardware shown in Fig. 1.25 that implements the 8-bit binary-to-BCD converter. Each of the five times through the *for* loop one of the rows of C_i components in Fig. 1.25 gets created. The identical C_i components in Fig. 1.25 implement the *if* statements within the *for* loop.

The 8-bit binary-to-BCD converter in Listing 1.14 can be extended to a 14-bit binary-to-BCD converter as shown in Listing 1.15.

Listing 1.14 binbcd8.vhd

```vhdl
-- Example 12a: 8-Bit Binary-to-BCD Converter
library IEEE;
use IEEE.STD_LOGIC_1164.all;
use IEEE.std_logic_unsigned.all;

entity binbcd8 is
  port(
        b : in STD_LOGIC_VECTOR(7 downto 0);
        p : out STD_LOGIC_VECTOR(9 downto 0)
      );
end binbcd8;

architecture binbcd8 of binbcd8 is
begin
  bcd1: process(b)

  variable z: STD_LOGIC_VECTOR (17 downto 0);

  begin
        for i in 0 to 17 loop
            z(i) := '0';
        end loop;
        z(10 downto 3) := b;
        for i in 0 to 4 loop
            if z(11 downto 8) > 4 then
                z(11 downto 8) := z(11 downto 8) + 3;
            end if;
            if z(15 downto 12) > 4 then
                z(15 downto 12) := z(15 downto 12) + 3;
            end if;
            z(17 downto 1) := z(16 downto 0);
        end loop;
        p <= z(17 downto 8);
  end process bcd1;
end binbcd8;
```

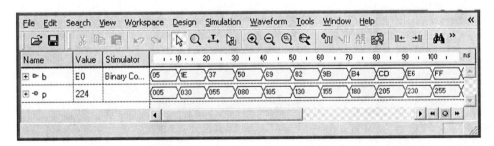

Figure 1.24 Simulation of the VHDL program in Listing 1.14

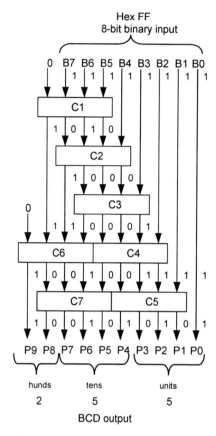

Figure 1.25 Hardware generated by the VHDL program in Listing 1.14

Listing 1.15 binbcd14.vhd

```vhdl
-- Example 12b:   14-bit Binary-to-BCD Converter
library IEEE;
use IEEE.std_logic_1164.all;
use IEEE.std_logic_unsigned.all;

entity binbcd14 is
    port (
        b: in STD_LOGIC_VECTOR (13 downto 0);
        p: out STD_LOGIC_VECTOR (16 downto 0)
    );
end binbcd14;

architecture binbcd14_arch of binbcd14 is
begin
  bcd1: process(b)
  variable z: STD_LOGIC_VECTOR (32 downto 0);
  begin
    for i in 0 to 32 loop
            z(i) := '0';
    end loop;
       z(16 downto 3) := B;
```

Listing 1.15 (cont.) binbcd14.vhd

```
      for i in 0 to 10 loop
          if z(17 downto 14) > 4 then
              z(17 downto 14) := z(17 downto 14) + 3;
          end if;
          if z(21 downto 18) > 4 then
              z(21 downto 18) := z(21 downto 18) + 3;
          end if;
          if z(25 downto 22) > 4 then
              z(25 downto 22) := z(25 downto 22) + 3;
          end if;
          if z(29 downto 26) > 4 then
              z(29 downto 26) := z(29 downto 26) + 3;
          end if;
          z(32 downto 1) := z(31 downto 0);
      end loop;
          p <= z(30 downto 14);
  end process bcd1;
end binbcd14_arch;
```

Example 13: Gray Code Converter

In this example we will show how to design a binary to Gray code converter and a Gray code to binary converter using VHDL.

Gray Code

A Gray code is an ordering of 2^n binary numbers such that only one bit changes from one entry to the next. The following shows the relationship between a 3-bit binary coding and a 3-bit Gray coding.

```
Binary code {0..7}: {000, 001, 010, 011, 100, 101, 110, 111}
Gray code   {0..7}: {000, 001, 011, 010, 110, 111, 101, 100}
```

Such a Gray code is not unique. One method for generating a Gray code sequence is to start with all bits zero and successively flip the right-most bit that produces a new string. Note that this works for the Gray coding shown above. The following algorithms can be used to convert a binary string, $b(i)$, $i = n-1 : 0$, to a Gray code string, $g(i)$, $i = n-1 : 0$, and vice versa.

Convert Binary to Gray:
    Copy the most significant bit.
    For each smaller *i*
    $g(i) = b(i+1) \wedge b(i)$

Convert Gray to Binary:
    Copy the most significant bit.
    For each smaller *i*
    $b(i) = b(i+1) \wedge g(i)$

4-Bit Binary to Gray Code Converter

Listing 1.16 is a VHDL program called *bin2gray* that will convert a 4-bit binary number $b(3:0)$ to a 4-bit Gray code $g(3:0)$ by following the algorithm given above. A simulation of this program is shown in Fig. 1.26.

Listing 1.16 bin2gray.vhd

```vhdl
// Example 13a: Binary to Gray code converter
library IEEE;
use IEEE.STD_LOGIC_1164.all;

entity bin2gray is
      port(
            b : in STD_LOGIC_VECTOR(3 downto 0);
            g : out STD_LOGIC_VECTOR(3 downto 0)
          );
end bin2gray;

architecture bin2gray of bin2gray is
begin
      process(b)
      begin
            g(3) <= b(3);
            for i in 2 downto 0 loop
                  g(i) <= b(i+1) xor b(i);
            end loop;
      end process;
end bin2gray;
```

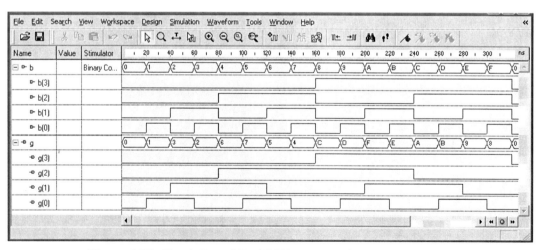

Figure 1.26 Simulation of the VHDL program in Listing 1.16

4-Bit Gray Code to Binary Converter

Listing 1.17 is a VHDL program called *gray2bin* that will convert a 4-bit Gray code $g(3:0)$ to a 4-bit binary number $b(3:0)$ by following the algorithm given above. Note in this case it is necessary to use the variable *bv* in the *for* loop because the value of $bv(i)$ depends on the value of $bv(i+1)$ and therefore these values need to be computed at the time the statement is executed. Recall that signal assignment statements are only executed at the end of the process. A simulation of this program is shown in Fig. 1.27.

Listing 1.17 gray2bin.vhd

```vhdl
// Example 13b: Gray code to binary converter
library IEEE;
use IEEE.STD_LOGIC_1164.all;

entity gray2bin is
    port(
        g : in STD_LOGIC_VECTOR(3 downto 0);
        b : out STD_LOGIC_VECTOR(3 downto 0)
    );
end gray2bin;

architecture gray2bin of gray2bin is
begin
    process(g)
    variable bv: STD_LOGIC_VECTOR(3 downto 0);
    begin
        bv(3) := g(3);
        for i in 2 downto 0 loop
            bv(i) := bv(i+1) xor g(i);
        end loop;
        b <= bv;
    end process;
end gray2bin;
```

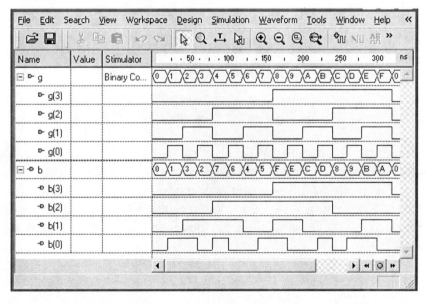

Figure 1.27 Simulation of the Verilog program in Listing 1.17

Example 14: Multiplier

In this example we will show how to design a multiplier using VHDL. The binary multiplication shown in Fig. 1.28 includes the intermediate additions of the partial products. This multiplication can be implemented directly in VHDL by using a *for* loop as shown in Listing 1.18. Recall from Example 11 that a VHDL *for* loop will "unwrap" to produce a combinational logic circuit. In Listing 1.18 the four passes through the *for* loop will produce the equivalent of the four cascaded components shown in the logic diagram in Fig. 1.29. Note that each component contains an adder, a multiplexer, and a shifter. A simulation of Listing 1.18 is shown in Fig. 1.30.

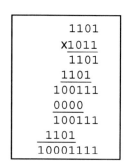

Figure 1.28
Binary multiplication

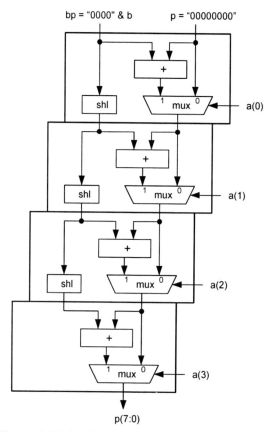

Figure 1.29 Logic diagram of Listing 1.18

It would be convenient to be able to make a 4-bit multiplier (or any size multiplier) by just using a * sign in a VHDL statement. In fact, we can! When you write a * b in a VHDL program the compiler will produce a combinational multiplier of the type we designed in Listing 1.18. For example, the VHDL program shown in Listing 1.19 will produce the same simulation as shown in Fig. 1.30. You must include the library statement

```
use IEEE.STD_LOGIC_unsigned.all;
```

in order to use the multiplication operator *.

You need to remember that using the simple * operator will generate a lot of hardware to produce the equivalent of the logic diagram in Fig. 1.29. Each additional bit of the *a* operand will add another component containing an adder, multiplexer, and shifter.

Listing 1.18 mult4a.vhd

```vhdl
-- Example 14a: 4-bit multiplier
library IEEE;
use IEEE.STD_LOGIC_1164.all;
use IEEE.STD_LOGIC_unsigned.all;

entity mult4a is
        port(
                a : in STD_LOGIC_VECTOR(3 downto 0);
                b : in STD_LOGIC_VECTOR(3 downto 0);
                p : out STD_LOGIC_VECTOR(7 downto 0)
            );
end mult4a;

architecture mult4a of mult4a is
begin
        process(a,b)
        variable pv,bp: STD_LOGIC_VECTOR(7 downto 0);
        begin
                pv := "00000000";
                bp := "0000" & b;
                for i in 0 to 3 loop
                        if a(i) = '1' then
                                pv := pv + bp;
                        end if;
                        bp := bp(6 downto 0) & '0';
                end loop;
                p <= pv;
        end process;
end mult4a;
```

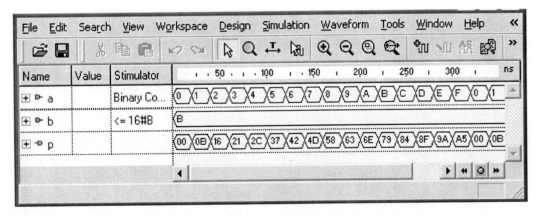

Figure 1.30 Simulation of the VHDL program in Listing 1.18

Listing 1.19 mult4b.vhd

```vhdl
-- Example 14b: 4-bit multiplier
library IEEE;
use IEEE.STD_LOGIC_1164.all;
use IEEE.STD_LOGIC_unsigned.all;

entity mult4b is
        port(
                a : in STD_LOGIC_VECTOR(3 downto 0);
                b : in STD_LOGIC_VECTOR(3 downto 0);
                p : out STD_LOGIC_VECTOR(7 downto 0)
            );
end mult4b;

architecture mult4b of mult4b is
begin
            p <= a * b;
end mult4b;
```

A Multiplication Function

A VHDL function returns a single value and can be used to help modularize your programs. Listing 1.20 shows how the multiplication algorithm that we implemented as a component in Listing 1.18 can be implemented as a function.

Listing 1.20 mult4c.vhd

```vhdl
-- Example 14c: 4-bit multiplier function
library IEEE;
use IEEE.STD_LOGIC_1164.all;
use IEEE.STD_LOGIC_unsigned.all;

entity mult4c is
        port(
                sw : in STD_LOGIC_VECTOR(7 downto 0);
                ld : out STD_LOGIC_VECTOR(7 downto 0)
            );
end mult4c;

architecture mult4c of mult4c is
function mul(a,b : in STD_LOGIC_VECTOR)
                                return STD_LOGIC_VECTOR is
variable pv,bp: STD_LOGIC_VECTOR(7 downto 0);
begin
        pv := "00000000";
        bp := "0000" & b;
        for i in 0 to 3 loop
                if a(i) = '1' then
                        pv := pv + bp;
                end if;
                bp := bp(6 downto 0) & '0';
        end loop;
        return pv;
end function;
```

Listing 1.20 (cont.) mult4c.vhd

```vhdl
signal a1,a2: STD_LOGIC_VECTOR(3 downto 0);
begin
      a1 <= sw(7 downto 4);
      a2 <= sw(3 downto 0);
      ld <= mul(a1,a2);

end mult4c;
```

The entity *mult4c* in Listing 1.20 calls the function *mul* and multiplies the 4-bit value of *sw*(7:4) by the 4-bit value of *sw*(3:0) and stores the product in *ld*(7:0). A simulation of Listing 1.20 is shown in Fig. 1.31. Note that the upper and lower nibble of *sw* are multiplied together to produce the product in *ld*.

The multiplication operator * shown in Listing 1.19 is defined in the *std_logic_arith* package (called in *unsigned.vhd*) using a function very similar to that in Listing 1.20.

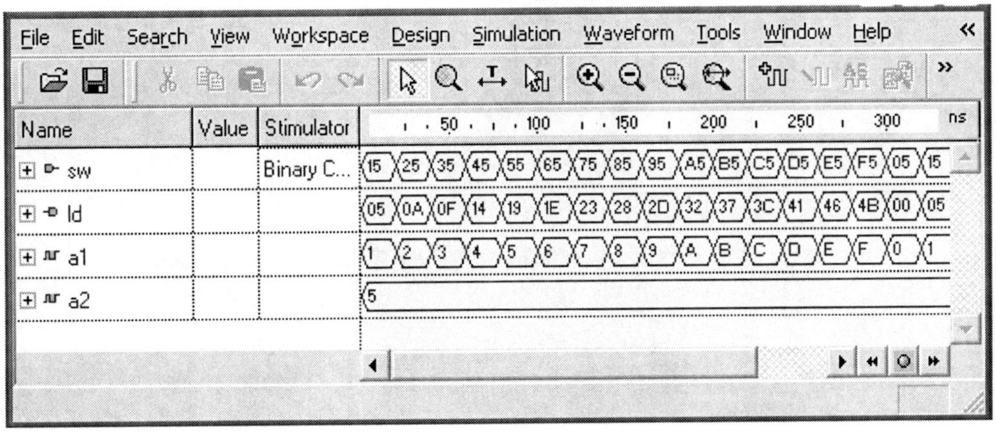

Figure 1.31 Simulation of the VHDL program in Listing 1.20

Example 15: Divider

In this example we will show how to design a divider using VHDL. Binary division can be carried out by long division as shown on the left in Fig. 1.32. This is equivalent to dividing the hex number 87 by D and gives the quotient A and a remainder 5.

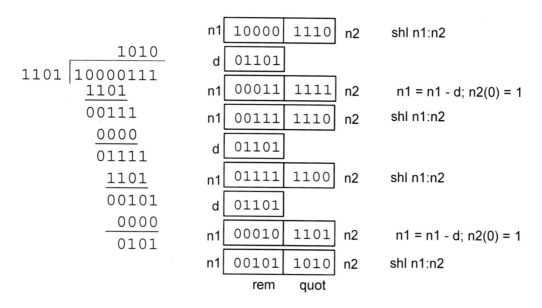

Figure 1.32 Example of binary division

The right-hand side of Fig. 1.32 shows how this division can be carried out by using the following algorithm:

1. Store the numerator in the concatenation of *n1*:*n2*
2. Store the denominator in *d*
3. Repeat 4 times:
   Shift *n1*:*n2* left one bit
   If *n1* > *d*
       n1 = *n1* – *d*;
       n2(0) = 1;
4. *quot* = *n2*;
   rem = *n1*(3:0);

An implementation of this algorithm in VHDL is shown in Listing 1.21 and its simulation is shown in Fig. 1.32.

Listing 1.21 div4.vhd

```vhdl
-- Example 15: 4-bit divider
library IEEE;
use IEEE.STD_LOGIC_1164.all;
use IEEE.STD_LOGIC_unsigned.all;

entity div4 is
        port(
              numer : in STD_LOGIC_VECTOR(7 downto 0);
              denom : in STD_LOGIC_VECTOR(3 downto 0);
              quotient : out STD_LOGIC_VECTOR(3 downto 0);
              remainder : out STD_LOGIC_VECTOR(3 downto 0)
            );
end div4;

architecture div4 of div4 is
begin
        process(numer,denom)
        variable d,n1: STD_LOGIC_VECTOR(4 downto 0);
        variable n2: STD_LOGIC_VECTOR(3 downto 0);
        begin
             d := '0' & denom;
             n2 := numer(3 downto 0);
             n1 := '0' & numer(7 downto 4);
             for i in 0 to 3 loop
                  n1 := n1(3 downto 0) & n2(3);
                  n2 := n2(2 downto 0) & '0';
                  if n1 >= d then
                        n1 := n1 - d;
                        n2(0) := '1';
                  end if;
             end loop;
             quotient <= n2;
             remainder <= n1(3 downto 0);
        end process;
end div4;
```

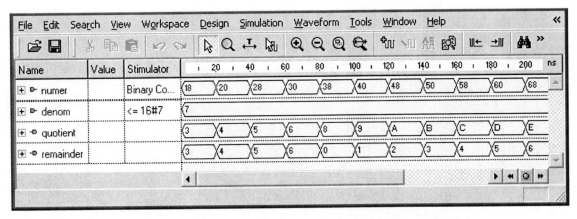

Figure 1.32 Simulation of the VHDL program in Listing 1.21

Example 16: 3-to-8 Decoder

In this example we will show how to design a 3-to-8 decoder in VHDL by using a *for* loop. The truth table for a 3-to-8 decoder is shown in Fig. 1.33. Note that the logic equation for each output, y_i, is just the minterm, m_i. You could implement this decoder in VHDL by writing the minterm logic equations or by using a *case* statement or a *with...select...then* statement to directly implement the truth table in Fig. 1.33. Alternatively, you could describe the behavior of this decoder using the VHDL program given in Listing 1.22. The simulation is shown in Fig. 1.34.

a2 a1 a0	y0 y1 y2 y3 y4 y5 y6 y7
0 0 0	1 0 0 0 0 0 0 0
0 0 1	0 1 0 0 0 0 0 0
0 1 0	0 0 1 0 0 0 0 0
0 1 1	0 0 0 1 0 0 0 0
1 0 0	0 0 0 0 1 0 0 0
1 0 1	0 0 0 0 0 1 0 0
1 1 0	0 0 0 0 0 0 1 0
1 1 1	0 0 0 0 0 0 0 1

Figure 1.33 Truth table for a 3-to-8 decoder

Listing 1.22 decoder38.vhd

```
-- Example 16: 3-to-8 decoder
library IEEE;
use IEEE.STD_LOGIC_1164.all;
use IEEE.STD_LOGIC_arith.all;
use IEEE.STD_LOGIC_unsigned.all;

entity decoder38 is
        port(
                a : in STD_LOGIC_VECTOR(2 downto 0);
                y : out STD_LOGIC_VECTOR(0 to 7)
            );
end decoder38;

architecture decoder38 of decoder38 is
begin
  process(a)
  variable j: integer;
  begin
      j := conv_integer(a);
      for i in 0 to 7 loop
         if(i = j) then
            y(i) <= '1';
         else
            y(i) <= '0';
         end if;
      end loop;
   end process;
end decoder38;
```

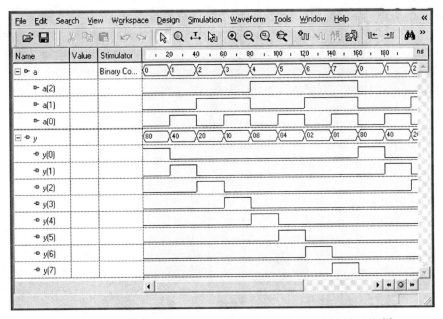

Figure 1.34 Simulation of the VHDL program in Listing 1.22

1.5 State Machines

In this section we will show how to implement a state machine in VHDL by using the examples of a sequence detector, a door lock code, and a traffic light controller.

Mealy and Moore State Machines

A canonical sequential network can be described by the diagram shown in Fig. 1.35. The present input is described by $x(t)$ and the present output is described by $z(t)$. Note that the present state $s(t)$ is the output of a state register, i.e., a series of flip-flops. The combinational network in Fig. 1.35 has two outputs: the output $z(t)$ and the next state $s(t+1)$.

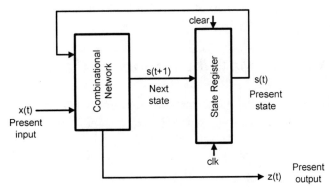

Figure 1.35 A canonical sequential network

It is often convenient to split the combinational network in Fig. 1.35 into two parts, C1 and C2, as shown in Fig. 1.36. The combinational component C1 has the present input $x(t)$ and the present state $s(t)$ as inputs, and outputs the next state $s(t+1)$. The combinational component C2 has the present input $x(t)$ and the present state $s(t)$ as inputs, and outputs the present output $z(t)$. This state machine in which the present output $z(t)$ depends on both the present state $s(t)$ and the present input $x(t)$ is called a *Mealy machine*.

If the present output $z(t)$ depends on only the present state $s(t)$ as shown in Fig. 1.37, we refer to the state machine as a *Moore machine*.

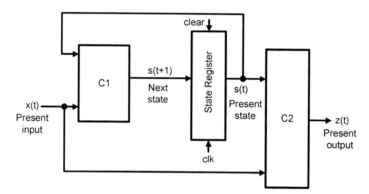

Figure 1.36 A Mealy state machine

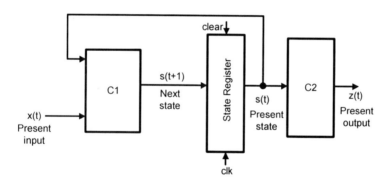

Figure 1.37 A Moore state machine

Example 17: A Moore Machine Sequence Detector

In this example we will show how to design a Moore machine to detect the sequence 1101. In the next example we will use a Mealy machine to detect the same sequence.

The state diagram for detecting the sequence 1101 using a Moore machine is shown in Fig. 1.38. The output, z, of this Moore machine will be 1 when the sequence 1101 is detected. We will let state $s0$ be the initial state in which no 1's have been received. Inasmuch as this is a Moore machine in which the output depends only on the present state, we will display the output within the circle in the state diagram under the state name as shown in Fig. 1.38. If we are in state $s0$ and the input is 0 we will stay in state *s0* as indicated by the arc in Fig. 1.38 where we have written the input next to the arc.

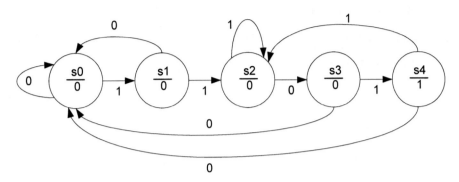

Figure 1.38 State diagram for detecting the sequence 1101

If the input is 1 then we move to state *s1* as shown in Fig. 1.38, which indicates that a single 1 has been received. If we are in state *s1* and the input is 0 we must move back to state *s0*. On the other hand, if we are in state *s1* and the input is 1 then we move to state *s2*, which indicates that two consecutive 1's have been received. If we are in state *s2* and the input is 1 we will stay in state *s2*. On the other hand, if we are in state *s2* and the input is 0 then we move to state s3, which indicates that the sequence 110 has been received. At this point if we receive an input 0 we must go all the way back to state *s0*.

Finally, if we are in state *s3* and the input is 1 then we move to state *s4* as shown in Fig. 1.38, which indicates that the sequence 1101 has been received. We therefore set the output to 1 in state *s4*. Note that if we are in state *s4* and the input is 0 then we must go all the way back to state *s0*. On the other hand, if we are in state *s4* and the input is 1 then we move back to state *s2*, which indicates that two consecutive 1's have been received. This is the transition that allows for overlapping bits to be used. The last 1 in the sequence detected, 1101, can be used as the first 1 in the next sequence being sought.

If we did not want to allow the reuse of the last bit for the possible first bit of the next sequence, the state machine would simply be modified to transition from *s4* to *s1* if a 1 were received as the next input. That is, the state machine will have received a single 1 towards the next sequence instead of two.

The VHDL program shown in Listing 1.23 implements this state diagram by following the basic Moore machine model shown in Fig. 1.37. Note the use of the enumerated type called *state_type* in Listing 1.23 to define the five states by their names *s0*, *s1*, *s2*, *s3*, and *s4*. The first process in Listing 1.23 implements the state register in Fig. 1.37. The second process in Listing 1.23 implements the combinational block *C1* in Fig. 1.37, and the third process in Listing 1.23 implements the combinational block *C2* in Fig. 1.37. This basic VHDL structure made from three processes can be used to implement any Moore machine. The one sequential block defines the state register and can normally be identical to the first process in Listing 1.23.

Listing 1.23 seqdeta.vhd

```vhdl
-- Example 17: Detect 1101 with Moore machine
library IEEE;
use IEEE.STD_LOGIC_1164.all;

entity seqdeta is
  port (clk: in STD_LOGIC;
        clr: in STD_LOGIC;
        din: in STD_LOGIC;
        dout: out STD_LOGIC);
end seqdeta;

architecture seqdeta of seqdeta is
type state_type is (s0, s1, s2, s3, s4);
signal present_state, next_state: state_type;
begin
sreg: process(clk, clr)
begin
     if clr = '1' then
            present_state <= s0;
     elsif clk'event and clk = '1' then
            present_state <= next_state;
     end if;
end process;


C1: process(present_state, din)
begin
  case present_state is
      when s0 =>
        if din = '1' then
          next_state <= s1;
        else
          next_state <= s0;
        end if;
      when s1 =>
        if din = '1' then
          next_state <= s2;
        else
          next_state <= s0;
        end if;
      when s2 =>
        if din = '0' then
          next_state <= s3;
        else
          next_state <= s2;
        end if;
      when s3 =>
        if din = '1' then
          next_state <= s4;
        else
          next_state <= s0;
        end if;
      when s4 =>
        if din = '0' then
          next_state <= s0;
        else
          next_state <= s2;
        end if;
      when others =>
        null;
  end case;
end process;
```

Listing 1.23 (cont.) seqdeta.vhd

```
C2: process(present_state)
begin
  if present_state = s4 then
    dout <= '1';
  else
    dout <= '0';
  end if;
end process;

end seqdeta;
```

The combinational block *C1* in Listing 1.23 is used to find the next state and normally contains a *case* statement that directly follows the state diagram (see Fig. 1.38). Finally, the combinational block *C2* defines the outputs, which for a Moore machine depend only on the current state. Normally, either an *if* statement or a *case* statement is used in the *C2* process to define the outputs. A simulation of Listing 1.23 is shown in Fig. 1.39.

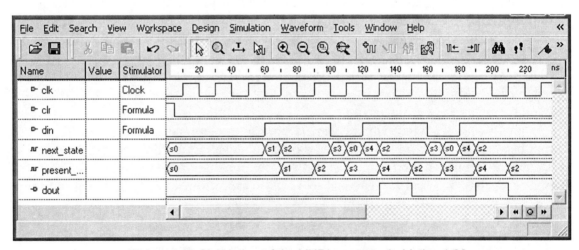

Figure 1.39 Simulation of the VHDL program in Listing 1.23

Example 18: A Mealy Machine Sequence Detector

The Moore machine state diagram in Fig. 1.38 has five states. The output, *z*, in this Moore machine is 1 when the state is *s4*. Another way to implement this sequence detector is to use a Mealy machine where the output, *z*, is 1 when the state is *s3* and *din* is 1. The state diagram for this Mealy machine used to implement the sequence detector is shown in Fig. 1.40. Note that it has only four states.

In Fig. 1.40 the outputs are shown on the transitions along with the input conditions. Transitions from the *present state* are labeled with "*present_input / present_output*". It is important to understand that the output value displayed is the combinational output that depends on the present input and the present state as shown in Fig. 1.36. For example, when the state is *s3* (meaning a 110 has been received) and the input becomes 1, the output *z* will become 1. On the next rising edge of the clock the state will change to s1 and the output z will change to 0. This means that the output z will never be latched to 1!

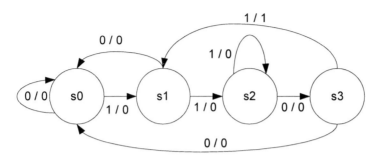

Figure 1.40 State diagram for detecting the sequence 1101 with a Mealy machine

If we want z to be a registered output (so it will keep its value when we transition to state s1) we can add a flip-flop to the output z(t) in Fig. 1.36. That is, the output of the combinational component C2 in Fig. 1.36 will be connected to the D-input of a D flip-flop. Thus, if in Fig. 1.40 the state is s3 and the input becomes 1, the output z will be 1, and on the next rising edge of the clock this output value of 1 will be latched into the output flip-flop, and the state will change to s1.

An equivalent way to implement this in VHDL is to make the output combinational component C2 in Fig. 1.36 a sequential component *Seq2* as shown in Listing 1.24. Note that the way to do this is to use the *clk'event and clk* = '1' phrase in an *if* statement within the process in such a way that the output *dout* gets assigned a value only on the rising edge of the clock.

In Listing 1.24, the output, *dout*, is no longer combinational but is a registered output because it occurs on the left-hand side of an equation within a *clk'event and clk* = '1' statement. Thus, on the rising edge of the clock the output *dout* will be latched to 1 if the present state is s3 and *din* is equal to 1. Also note from Listing 1.24 that only four states are necessary to implement the Mealy machine as shown in Fig. 1.40. A simulation of the VHDL program in Listing 1.24 is shown in Fig. 1.41.

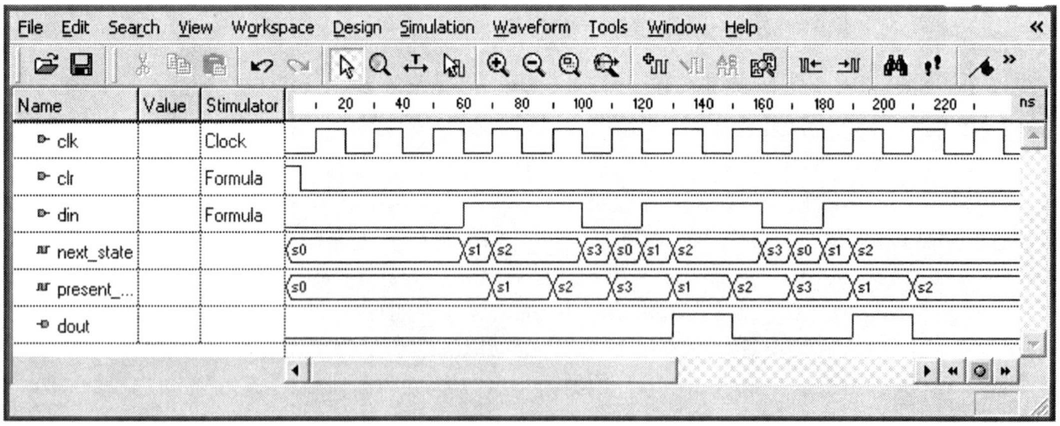

Figure 1.41 Simulation of the VHDL program in Listing 1.24

Listing 1.24 seqdetb.vhd

```vhdl
-- Example 18: Detect 1101 with Mealy machine
library IEEE;
use IEEE.STD_LOGIC_1164.all;

entity seqdetb is
  port (clk: in STD_LOGIC;
        clr: in STD_LOGIC;
        din: in STD_LOGIC;
        dout: out STD_LOGIC);
end seqdetb;

architecture seqdetb of seqdetb is
type state_type is (s0, s1, s2, s3);
signal present_state, next_state: state_type;
begin
sreg: process(clk, clr)
begin
      if clr = '1' then
           present_state <= s0;
      elsif clk'event and clk = '1' then
           present_state <= next_state;
      end if;
end process;

C1: process(present_state, din)
begin
  case present_state is
      when s0 =>
         if din = '1' then
            next_state <= s1;
         else
            next_state <= s0;
         end if;
      when s1 =>
         if din = '1' then
            next_state <= s2;
         else
            next_state <= s0;
         end if;
      when s2 =>
         if din = '0' then
            next_state <= s3;
         else
            next_state <= s2;
         end if;
      when s3 =>
         if din = '1' then
            next_state <= s1;
         else
            next_state <= s0;
         end if;
      when others =>
         null;
  end case;
end process;
```

Listing 1.24 (cont.) seqdetb.vhd

```
Seq2: process(clk, clr)
begin
     if clr = '1' then
          dout <= '0';
     elsif clk'event and clk = '1' then
          if present_state = s3 and din = '1' then
               dout <= '1';
          else
               dout <= '0';
          end if;
     end if;
end process;

end seqdetb;
```

Example 19: Door Lock Code

In this example we will expand the sequence detector in Example 17 to simulate a door lock involving three buttons in which you press a 4-digit code. We will use *btn*(2:0) on the Nexys-2 board for the door lock buttons. For example, you might press buttons 2-0-1-2 to open the door. We will make the code programmable using the switches according to the following scheme:

- The *first* correct button selected from *btn*(2:0) will be the switch setting *sw*(7:6) where valid settings are "00", "01", "10" corresponding to *btn*(0), *btn*(1), and *btn*(2) respectively.
- The *second* correct button selected from *btn*(2:0) will be the switch setting *sw*(5:4).
- The *third* correct button selected from *btn*(2:0) will be the switch setting *sw*(3:2).
- The *fourth* correct button selected from *btn*(2:0) will be the switch setting *sw*(1:0).

You must make four button pressings before you know if you were successful. If you entered the correct code the LED *ld*(1) will light up. If you enter an incorrect code the LED *ld*(0) will light up.

The top-level design is shown in Fig. 1.42. Note that you will generate a clock pulse if you press any of the three pushbuttons *btn*(0), *btn*(1), or *btn*(2). The 2-bit input *bn*(1:0) to the doorlock component will be 00 if you press *btn*(0), 01 if you press *btn*(1), and 10 if you press *btn*(2). These values can be compared to the switch settings to produce the state diagram for doorlock shown in Fig. 1.43. Note that any time you press an incorrect

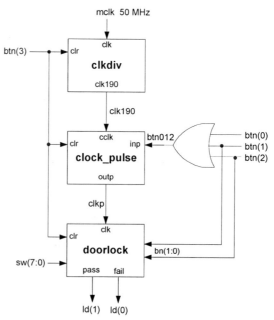

Figure 1.42 Top-level design for doorlock code

button you must continue to press a total of four buttons before entering the *fail* state *er*4.

The VHDL program for this *doorlock* component is given in Listing 1.25 and its simulation is shown in Fig. 1.44. The VHDL program for the top-level design in Fig. 1.42 is given in Listing 1.26.

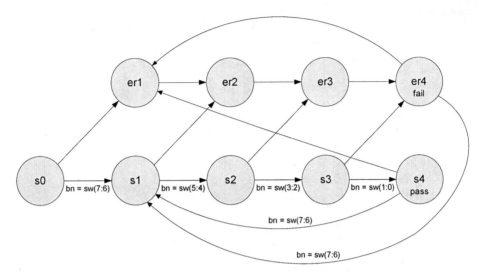

Figure 1.43 State diagram for doorlock code

Listing 1.25 doorlock.vhd

```vhdl
-- Example 19a: Detect doorlock code from switch settings
library IEEE;
use IEEE.STD_LOGIC_1164.all;

entity doorlock is
  port (clk: in STD_LOGIC;
        clr: in STD_LOGIC;
        bn: in STD_LOGIC_VECTOR(1 downto 0);
        sw: in STD_LOGIC_VECTOR(7 downto 0);
        pass: out STD_LOGIC;
        fail: out STD_LOGIC);
end doorlock;

architecture doorlock of doorlock is

type state_type is (s0,s1,s2,s3,s4,er1,er2,er3,er4);
signal present_state, next_state: state_type;

begin

sreg: process(clk, clr)
begin
     if clr = '1' then
            present_state <= s0;
     elsif clk'event and clk = '1' then
            present_state <= next_state;
       end if;
end process;
```

Listing 1.25 (cont.) doorlock.vhd

```vhdl
C1: process(present_state, bn)
begin
  case present_state is
      when s0 =>
        if bn = sw(7 downto 6) then
          next_state <= s1;
        else
          next_state <= er1;
        end if;
      when s1 =>
        if bn = sw(5 downto 4) then
          next_state <= s2;
        else
          next_state <= er2;
        end if;
      when s2 =>
        if bn = sw(3 downto 2) then
          next_state <= s3;
        else
          next_state <= er3;
        end if;
      when s3 =>
        if bn = sw(1 downto 0) then
          next_state <= s4;
        else
          next_state <= er4;
        end if;
      when S4 =>
        if bn = sw(7 downto 6) then
          next_state <= s1;
        else
          next_state <= er1;
        end if;
      when er1 =>
          next_state <= er2;
      when er2 =>
          next_state <= er3;
      when er3 =>
          next_state <= er4;
      when er4 =>
        if bn = sw(7 downto 6) then
          next_state <= s1;
        else
          next_state <= er1;
        end if;
      when others =>
        null;
  end case;
end process;
```

Listing 1.25 (cont.) doorlock.vhd

```vhdl
C2: process(present_state)
begin
  if present_state = s4 then
    pass <= '1';
  else
    pass <= '0';
  end if;
  if present_state = er4 then
    fail <= '1';
  else
    fail <= '0';
  end if;
end process;

end doorlock;
```

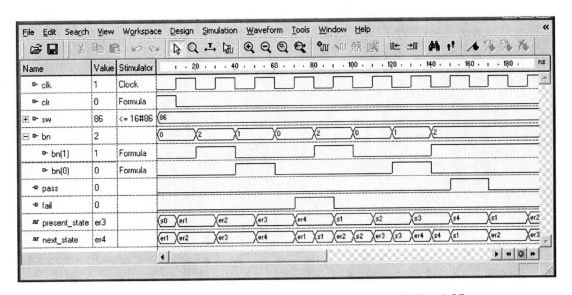

Figure 1.44 Simulation of the VHDL program in Listing 1.25

Listing 1.26 doorlock_top.vhd

```vhdl
-- Example 19b: Doorlock code from switch settings: Nexys-2
library IEEE;
use IEEE.STD_LOGIC_1164.all;

entity doorlock_top is
    port(
        mclk : in STD_LOGIC;
        btn : in STD_LOGIC_VECTOR(3 downto 0);
        sw : in STD_LOGIC_VECTOR(7 downto 0);
        ld : out STD_LOGIC_VECTOR(1 downto 0)
        );
end doorlock_top;

architecture doorlock_top of doorlock_top is

    component clock_pulse
    port(
        inp : in std_logic;
        cclk : in std_logic;
        clr : in std_logic;
        outp : out std_logic);
    end component;

    component clkdiv
    port(
        mclk : in std_logic;
        clr : in std_logic;
        clk190 : out std_logic);
    end component;

    component doorlock
    port(
        clk : in std_logic;
        clr : in std_logic;
        bn : in std_logic_vector(1 downto 0);
        sw : in std_logic_vector(7 downto 0);
        pass : out std_logic;
        fail : out std_logic);
    end component;

    signal clr, clk190, clkp, btn012: std_logic;
    signal bn : std_logic_vector(1 downto 0);
begin

    btn012 <= btn(2) or btn(1) or btn(0);
    clr <= btn(3);
    bn(1) <= btn(2);
    bn(0) <= btn(1);

U1 : clkdiv
    port map(
        mclk => mclk,
        clr => clr,
        clk190 => clk190
    );
```

Listing 1.26 (cont.) doorlock_top.vhd

```
U2 : clock_pulse
      port map(
            inp  => btn012,
            cclk => clk190,
            clr  => clr,
            outp => clkp
      );

U3 : doorlock
      port map(
            clk  => clkp,
            clr  => clr,
            bn   => bn,
            sw   => sw,
            pass => ld(1),
            fail => ld(0)
      );

end doorlock_top;
```

1.6 VHDL Package

A VHDL *package* provides a way of organizing your designs in a more convenient way. Packages are separate VHDL files that can be used to store component declarations, constants, other data types, and subprograms. As your programs get larger they will typically contain many *.vhd* files in the project. Each of these components (entities) must have its own component declaration in the architecture declaration portion of the top-level design. For example, in Listing 1.26 of Example 19 the three component declarations for *clock_pulse*, *clkdiv*, and *doorlock* are included. Larger programs might include a dozen or more such declarations, which might take up as much or more room in the architecture as the main part of the program. Luckily, there is a way of moving all of these component declarations to a package in a separate file. We will show how to do this for the door lock code program in Example 20.

In general, a package would contain a package declaration and an optional package body. For storing component decarations we only need a package declaration, whose general form is

                package <identifier> **is**
                    {package_declarative_item}
              end <identifier>;

A package that you have been using in all of your programs is the *std_logic_1164* package that is defined in the file *stdlogic.vhd* stored in the IEEE library. To access the contents of this package you have included the statement

```
use IEEE.STD_LOGIC_1164.all;
```

at the beginning of all of your programs. You will use a similar *use* clause to access packages that you define as we shall see in Example 20.

Example 20: Door Lock Code – Package

We can take the three component decarations from Listing 1.26 and put them in a VHDL package called *doorlock_components* as shown in Listing 1.27. This VHDL file called *doorlock_components.vhd* must be stored in your working directory and added to your project. Then your new top-level design must include the *use* clause

```
use work.doorlock_components.all;
```

at the beginning of your program as shown in Listing 1.28. This tells your program to find the package *doorlock_components* in your working directory and to use all of the component declarations in that package.

Listing 1.27 doorlock_components.vhd
```vhdl
-- A package containing component declarations
-- for door lock Example 20
library IEEE;
use IEEE.std_logic_1164.all;

package doorlock_components is

    component clock_pulse
    port(
        inp : in std_logic;
        cclk : in std_logic;
        clr : in std_logic;
        outp : out std_logic);
    end component;

    component clkdiv
    port(
        mclk : in std_logic;
        clr : in std_logic;
        clk190 : out std_logic);
    end component;

    component doorlock
    port(
        clk : in std_logic;
        clr : in std_logic;
        bn : in std_logic_vector(1 downto 0);
        sw : in std_logic_vector(7 downto 0);
        pass : out std_logic;
        fail : out std_logic);
    end component;

end doorlock_components;
```

Listing 1.28 doorlock2_top.vhd

```vhdl
-- Example 20: Doorlock code from switch settings: Nexys-2
library IEEE;
use IEEE.STD_LOGIC_1164.all;
use work.doorlock_components.all;

entity doorlock2_top is
        port(
                mclk : in STD_LOGIC;
                btn : in STD_LOGIC_VECTOR(3 downto 0);
                sw : in STD_LOGIC_VECTOR(7 downto 0);
                ld : out STD_LOGIC_VECTOR(1 downto 0)
            );
end doorlock2_top;

architecture doorlock2_top of doorlock2_top is
signal clr, clk190, clkp, btn012: std_logic;
signal bn : std_logic_vector(1 downto 0);
begin

        btn012 <= btn(2) or btn(1) or btn(0);
        clr <= btn(3);
        bn(1) <= btn(2);
        bn(0) <= btn(1);

U1 : clkdiv
        port map(
                mclk => mclk,
                clr => clr,
                clk190 => clk190
        );

U2 : clock_pulse
        port map(
                inp => btn012,
                cclk => clk190,
                clr => clr,
                outp => clkp
        );

U3 : doorlock
        port map(
                clk => clkp,
                clr => clr,
                bn => bn,
                sw => sw,
                pass => ld(1),
                fail => ld(0)
        );

end doorlock2_top;
```

1.7 Multiple-Process VHDL Programs

Instead of having a single .vhd file for each component, for which we must have a separate component declaration, we could simply write a separate process or other concurrent statement for each component within a single VHDL program. This has the advantage of having to add only that program to your project rather than all of the supporting .vhd files. The *x7segb* component to display hex numbers on the 7-segment display is a good example of this technique.

Example 21: The 7-Segment Display Module, *x7segb*

Each digit on the 7-segment display of the Nexys-2 board is enabled by one of the active low signals *an*(3:0) and all digits share the same *a_to_g*(6:0) signals. If *an*(3:0) = "0000" then all digits are enabled and display the same hex digit. To display a different hex digit on each display it is necessary to multiplex four different digit values to the *hex7seg* component and continually cycle through these four values quickly so that your eyes perceive the four digits at the same time. The logic circuit shown in Fig. 1.45 will do this.

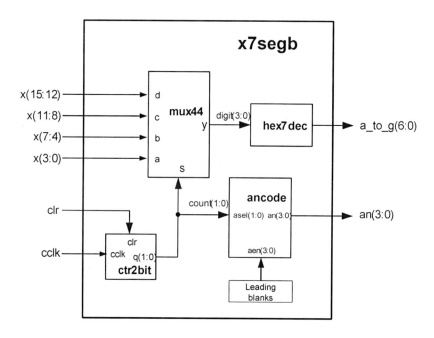

Figure 1.45 Digital circuit to multiplex 7-segment displays

The component *ctr2bit* is a 2-bit counter that cycles through the values 00, 01, 10, and 11 at a 190 Hz rate. This counter output is connected to the signal *count*(1:0), which is connected to the select input *s*(1:0) of the quad 4-to-1 mux and also to the *asel*(1:0) input of the component *ancode*.

The purpose of the component *ancode* is to enable one digit at a time depending on the value of the signal *count*(1:0). If *count* = "00" then the output *an*(3:0) will be "1110" enabling the rightmost 7-segment display *an0*. If *count* = "01" then the output *an*(3:0) will be "1101" enabling the *an*1 7-segment display. If *count* = "10" then the output *an*(3:0) will be "1011" enabling the *an*2 7-segment display. Finally if *count* = "11" then the output *an*(3:0) will be "0111" enabling the leftmost 7-segment display *an*3.

We could use the counter from Example 5, the multiplexer from Example 8, the 7-segment decoder from Example 9, and a separate file for the component *ancode*. We would then have to include component declarations for each of these components in the program for *x7segb*. It is more inconvenient to incorporate all of the components shown in Fig. 1.45 into a single file called *x7segb.vhd* as shown in Listing 1.29. This means that we will not have to load separate files for each component into a project every time we want to display a value on the 7-segment display. This program will display the 4-digit hex value *x*(15:0) on the four digits of the 7-segment display with leading blanks. Note that multiple processes within a single program execute concurrently.

Listing 1.29 x7segb.vhd

```vhdl
library IEEE;
use IEEE.STD_LOGIC_1164.all;
use IEEE.std_logic_unsigned.all;

entity x7segb is
    Port ( x : in std_logic_vector(15 downto 0);
           cclk, clr : in std_logic;
           a_to_g : out std_logic_vector(6 downto 0);
           an : out std_logic_vector(3 downto 0)
           );
end x7segb;

architecture arch_x7segb of x7segb is

    signal digit : std_logic_vector(3 downto 0);
    signal count : std_logic_vector(1 downto 0);
    signal aen: std_logic_vector(3 downto 0);

begin
    -- set aen(3:0) for leading blanks
       aen(3) <= (x(15) or x(14) or x(13) or x(12));
       aen(2) <= (x(15) or x(14) or x(13) or x(12) or x(11) or
                                    x(10) or x(9) or x(8));
       aen(1) <= (x(15) or x(14) or x(13) or x(12) or x(11) or
          x(10) or x(9) or x(8) or x(7) or x(6) or x(5) or x(4));
       aen(0) <= '1';  -- digit 0 always on

    -- 2-bit counter
    ctr2bit: process(cclk,clr)
        begin
            if(clr = '1') then
                count <= "00";
            elsif(cclk'event and cclk = '1') then
                count <= count + 1;
            end if;
        end process;
```

Listing 1.29 (cont.) x7segb.vhd

```vhdl
    -- MUX4
    with count select
        digit <=  x(3 downto 0)    when "00",
                  x(7 downto 4)    when "01",
                  x(11 downto 8)   when "10",
                  x(15 downto 12)  when others;

    -- seg7dec
    with digit select
    a_to_g <= "1001111" when "0001",  --1
              "0010010" when "0010",  --2
              "0000110" when "0011",  --3
              "1001100" when "0100",  --4
              "0100100" when "0101",  --5
              "0100000" when "0110",  --6
              "0001111" when "0111",  --7
              "0000000" when "1000",  --8
              "0000100" when "1001",  --9
              "0001000" when "1010",  --A
              "1100000" when "1011",  --b
              "0110001" when "1100",  --C
              "1000010" when "1101",  --d
              "0110000" when "1110",  --E
              "0111000" when "1111",  --F
              "0000001" when others;  --0

    -- digit select
    ancode: process(count)
        begin
            if(aen(conv_integer(count)) = '1') then
                an <= (others => '1');
                an(conv_integer(count)) <= '0';
            else
                an <= "1111";
            end if;
        end process;

end arch_x7segb;
```

The *ancode* component in Fig. 1.45 is implemented by the *ancode* process in Listing 1.29. The statement

```
an(conv_integer(count))
```

in Listing 1.29 will convert the value of *count*(1:0) of type STD_LOGIC_VECTOR to the corresponding integer values of 0, 1, 2, or 3 that can be used as the index of *an*. The statement

```
an <= (others => '1');
```

in Listing 1.29 will assign all values of *an*(3:0) to '1'. Thus, *an* will have the value "1111". The next statement

```
an(conv_integer(count)) <= '0';
```

will set one of these values to '0' depending on the value of *count*. Note that because these statements use the signal assignment statement <= the last value assigned within the process will be the one assigned at the end of the process. Thus, only one bit in *an* will be low thus enabling only one 7-segment display.

The simulation of this program is shown in Fig. 8.2.

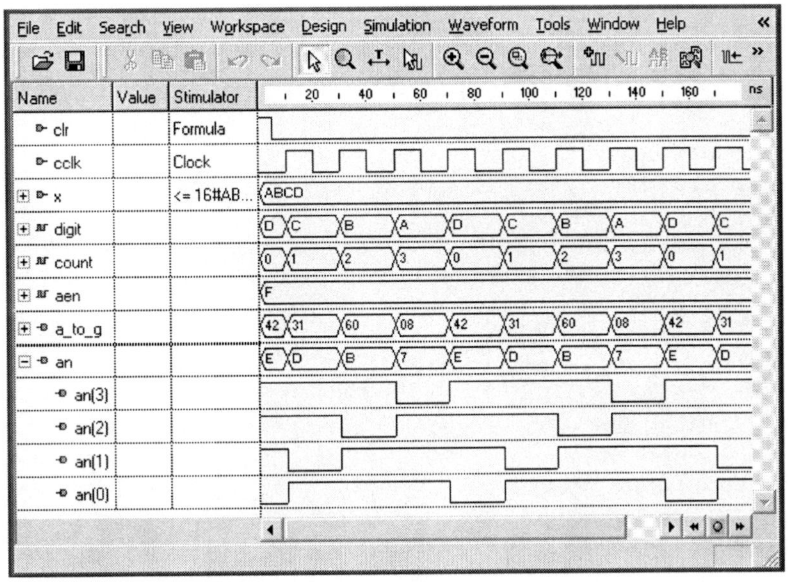

Figure 1.46 Simulation of the VHDL program in Listing 1.29

1.8 VHDL *while* Statement

All of the VHDL statement we have used in this chapter so far (including *if*, *case*, and *for.. loop*) can not only be simulated, but can also be synthesized to hardware. VHDL also has other statements including a *while* loop that can be simulated, but can *not* be synthesized to hardware. To see why this is the case consider Euclid's classic algorithm for finding the *greatest common divisor* (GCD) of two numbers, which is shown in Fig. 1.47.

```
Input: int x, y;
Output: int gcd;
while (x /= y) {
    if(x < y)
        y = y - x;
    else
        x = x - y;
}
gcd = x;
```

Figure 1.47 Euclid's GCD algrorithm

Example 22: GCD Algorithm – Part 1

Listing 1.30 shows how we might try to implement this algorithm directly in VHDL. If we simulate this program we get the correct answers as shown by the simulation in Fig. 1.48. But if we try to synthesize this program we get the error shown in Fig. 1.49. Why?

Listing 1.30 gcd1.vhd

```vhdl
-- Example 22: greatest common divisor
library IEEE;
use IEEE.STD_LOGIC_1164.all;
use IEEE.STD_LOGIC_unsigned.all;

entity gcd1 is
      port(
            x : in STD_LOGIC_VECTOR(3 downto 0);
            y : in STD_LOGIC_VECTOR(3 downto 0);
            gcd : out STD_LOGIC_VECTOR(3 downto 0)
          );
end gcd1;

architecture gcd1 of gcd1 is
begin
      process(x, y)
      variable xv, yv: STD_LOGIC_VECTOR(3 downto 0);
      begin
        xv := x;
        yv := y;
        while(xv /= yv) loop
          if xv < yv then
              yv := yv - xv;
          else
              xv := xv - yv;
          end if;
        end loop;
        gcd <= xv;
      end process;
end gcd1;
```

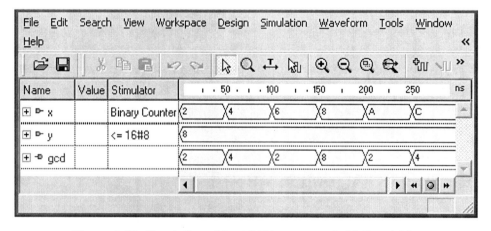

Figure 1.48 Simulation of the VHDL program in Listing 1.30

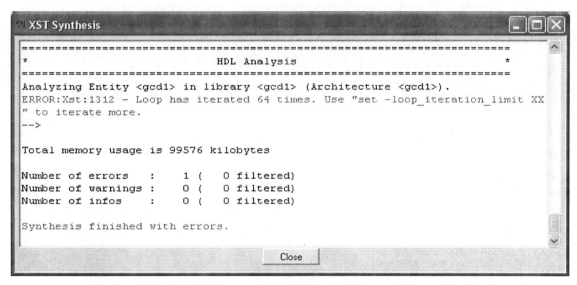

Figure 1.49 The GCD program in Listing 1.30 can *not* be synthesized

Note from the error in Fig. 1.49 that the synthesis program tried to iterate 64 times before giving up. Recall that when we synthesized *for* loops the circuits within the *for* loop "unwrapped" to produce a new instance of the circuit each time through the loop. But in *for* loops we know how many times the program goes through the loop ahead of time, so the synthesis tool knows how many instances of the circuit to create. However, in the *while* loop in Listing 1.30 the number of times the program goes through the loop depends on the input values x and y, which aren't known until the user enters these values long after the hardware has been created! This is why we can't use *while* loops in our VHDL programs if we want to create hardware on the FPGA.

Inasmuch as many of the algorithms we want to implement in an FPGA will be described using *while* loops, we must figure out how to implement such algorithms in hardware. Most of the rest of the book will be devoted to implementing such algorithms.

Problems

1.1. A 4-input OR gate can be implemented using three 2-input OR gates described by the VHDL statements

```
zv := x(1);
for i in 2 to 4 loop
    zv := zv or x(i);
end loop;
```

Write a VHDL program for this circuit and simulate it using Aldec Active-HDL. Choose a counter stimulator for X that counts from 0000 to 1111. Print out the resulting waveform. Draw the resulting logic diagram.

1.2. A 4-input XOR gate can be implemented using three 2-input XOR gates described by the VHDL statements

```
zv := x(1);
for i in 2 to 4 loop
    zv := zv xor x(i);
end loop;
```

Write a VHDL program for this circuit and simulate it using Aldec Active-HDL. Choose a counter stimulator for X that counts from 0000 to 1111. Print out the resulting waveform. Draw the resulting logic diagram.

1.3. Using DeMorgan's theorem a 4-input NOR gate can be implemented using three 2-input AND gates described by the VHDL statements

```
zv := not x(1);
for i in 2 to 4 loop
    zv := zv and not x(i);
end loop;
```

Write a VHDL program for this circuit and simulate it using Aldec Active-HDL. Choose a counter stimulator for X that counts from 0000 to 1111. Print out the resulting waveform. Draw the resulting logic diagram.

1.4. Using DeMorgan's theorem a 4-input NAND gate can be implemented using three 2-input OR gates described by the VHDL statements

```
zv := not x(1);
for i in 2 to 4 loop
    zv := zv or not x(i);
end loop;
```

Write a VHDL program for this circuit and simulate it using Aldec Active-HDL. Choose a counter stimulator for X that counts from 0000 to 1111. Print out the resulting waveform. Draw the resulting logic diagram.

1.5. A 4-input XNOR gate can be implemented using three 2-input XNOR gates described by the VHDL statements

```
zv := x(1);
for i in 2 to 4 loop
    zv := zv xnor x(i);
end loop;
```

Write a VHDL program for this circuit and simulate it using Aldec Active-HDL. Choose a counter stimulator for X that counts from 0000 to 1111. Print out the resulting waveform. Draw the resulting logic diagram.

1.6 Write a VHDL program for a 16-bit binary-to-BCD converter.

1.7 Write a VHDL program for a 9-bit binary-to-BCD converter.

1.8 Simulate the 14-bit binary-to-BCD converter given in Listing 1.15.

1.9 Draw the logic circuit that gets generated when the *for* loop in Listing 1.16 gets "unwrapped."

1.10 Draw the logic circuit that gets generated when the *for* loop in Listing 1.17 gets "unwrapped."

1.11 Write a VHDL program for a 5-bit binary to Gray code converter.

1.12 Write a VHDL program for a 5-bit Gray code to binary converter.

1.13 Write a VHDL program for a 6-bit binary to Gray code converter.

1.14 Write a VHDL program for a 6-bit Gray code to binary converter.

1.15 Write a VHDL program for a 8-bit binary to Gray code converter.

1.16 Write a VHDL program for a 8-bit Gray code to binary converter.

1.17 Write a VHDL program for an 8 x 8 multiplier using the *mul* function. Simulate the program using Active HDL.

1.18 Design a Moore machine to detect the sequence 1010. Design a Mealy machine to detect the same sequence.

1.19 Design a Moore machine to detect the sequence 1011. Design a Mealy machine to detect the same sequence.

1.20 Design a Moore machine to detect the sequence 10101. Design a Mealy machine to detect the same sequence.

2

Datapaths and Control Units

When trying to implement Euclid's GCD algorithm in Section 1.8 we discovered that a direct VHDL implementation of the algorithm would simulate, but would not synthesize. In this algorithm, which we show again in Fig. 2.1, it is the *while* loop that is the problem. There is no way of knowing, before the program is run, how many times to go through the *while* loop. When we gave x and y specific values in the simulation in Fig. 1.48 there is no problem and the *while* loop is "unwound" in a way similar to a *for* loop to produce a combinational circuit that implements the algorithm. But when the values of x and y are to be entered at run time, e.g., from the switch inputs, then the number of times to go through the *while* loop can't be known when the circuit is designed. This means that a combinational circuit won't work and we will need to design a sequential circuit in which the number of times through the *while* loop will vary, depending on the inputs when the program is run.

```
Input: int x, y;
Output: int gcd;
while (x /= y) {
    if(x < y)
        y = y - x;
    else
        x = x - y;
}
gcd = x;
```

Figure 2.1 Euclid's GCD algrorithm

In this chapter we will develop a general method that can be used to implement a digital processor that can implement any algorithm including one with a *while* loop. The general structure of a digital processor is shown in Fig. 2.2. The datapath contains registers, multiplexers, and various arithmetic and logical combinational logic components. The timing of the processor is handled by the control unit which typically contains a state machine. Various outputs from the datapath such as conditional flags are sent to the control unit. The control unit provides various control signals to the datapath such as register *load* signals and multiplexer *select* signals.

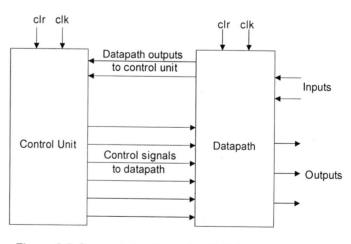

Figure 2.2 General structure of a digital processor

The following steps can be used to create a datapath for implementing an algorithm.

1. Draw a register (rectangular box with input at the top and output at the bottom) for each variable in the algorithm. For the algorithm in Fig. 2.1 these would include *x*, *y*, and *gcd*. Each register will also have a *clear*, *clock*, and *load* inputs. When the *clear* signal is high, the output of the register will be a predetermined initial value, normally zero. If the *load* signal is high, then on the next rising edge of the clock signal the input value will be loaded into the register and appear on the output.

2. Define combinational blocks to implement any necessary arithmetic or logical operation.

3. Connect the outputs of the registers to the inputs of the appropriate arithmetic and logical operations, and connect the outputs of the arithmetic and logical operations to the appropriate registers. Multiplexers can be used if the input to a register can come from more than one source.

Applying these steps to the GCD algorithm given in Fig. 2.1 results in the datapath shown in Fig. 2.3. The first step is to draw the three registers for the three variables *x*, *y*, and *gcd*. The next step is to draw two subtractors and two comparators and then connect them through the two multiplexers to implement all of the operations in the GCD algorithm.

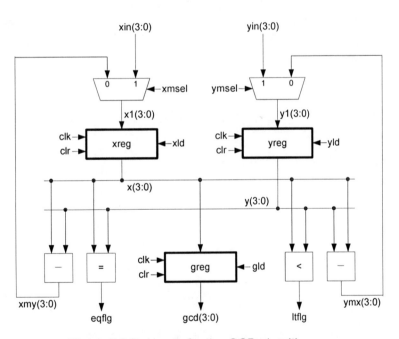

Figure 2.3 Datapath for the GCD algorithm

Note that the datapath only determines what gets stored in a particular register. There is no timing information in the datapath. The control unit provides all the timing information in the form of a *state machine*.

We will complete the GCD algorithm in Example 23 and then develop a datapath and control unit for a simple integer square root algorithm in Example 24.

Example 23

GCD Algorithm – Part 2

We saw above that the GCD algorithm given in Fig. 2.1 would have the datapath shown in Fig. 2.3. Note that this datapath has three registers, two multiplexers, two subtractors, and two logical comparators. We have shown the bus widths to be 4 bits in Fig. 2.3 so that we can implement this algorithm on the Nexys-2 board and use the switches for the inputs *xin* and *yin*. In Fig. 2.3 the inputs *clk*, *clr*, *xin*(3:0) and *yin*(3:0) would come from the outside, while the inputs *xmsel*, *ymsel*, *xld*, *yld*, and *gld* would come from the control unit as shown in Fig. 2.4. The outputs *eqflg* and *ltflg* from the datapath would go to the control unit, while the output *gcd*(3:0) goes to the output *gcd_out*(3:0) as shown in Fig. 2.4. The *go* input signal to the control unit might come from a pushbutton and would start the algorithm. The *clr* inputs might come from a pushbutton and the *xin*(3:0) and *yin*(3:0) inputs to the datapath might come from the slide switches. The output *gcd_out*(3:0) might go to the 7-segment display. The control unit will contain a state machine described by the state diagram in Fig. 2.5.

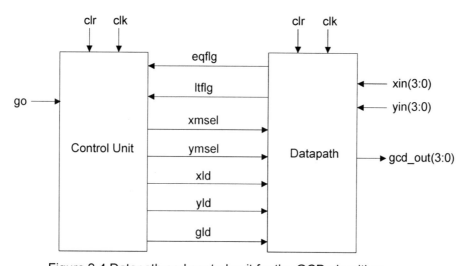

Figure 2.4 Datapath and control unit for the GCD algorithm

To draw the state machine in Fig. 2.5 you would simply follow the algorithm given in Fig. 2.1. Recall that the state machine will move from one state to the next on the rising edge of the clock. The algorithm will wait in the *start* state until the *go* signal goes high and then move to the *input* state. In this *input* state the signals *xmsel*, *ymsel*, *xld*, and *yld* will be set to 1 so that the two inputs *xin* and *yin* will be loaded into the registers *xreg* and *yreg* on the next rising edge of the clock. The state diagram in Fig. 2.5 then moves to the *test1* state. In this state the value of the datapath output *eqflg* will determine which state to go to next. If *eqflg* = '1' that means that $x = y$ and the *while* loop in Fig. 2.1 will be exited which means that the next state will be the *done* state in Fig. 2.5. Otherwise the state diagram moves to the *test2* state which will test to see if $x < y$. If it is then the datapath output *ltflg* will be 1 which will tell the control unit to move to state *update1* in Fig. 2.5, which from the algorithm in Fig. 2.1 means that you want to compute $y = y - x$. You would do this in the datapath by setting *yld* to 1 with *ymsel* = 0 so that the value of $y - x$ will be loaded into the *y* register on the next rising edge of the clock. This will take you back to state *test1* which is the beginning of the *while* loop in Fig. 2.1 to text again for the equality of *x* and *y*. When the algorithm is in state *test2* and the value of *ltflg* is zero it means that *x* is not less than *y* so the next state will be *update2* in which you want to compute $x = x - y$. You would do this in the datapath by setting *xld* to 1 with *xmsel* = 0 so that the value of $x - y$ will be loaded into the *x* register on the next rising edge of the clock. When the algorithm gets to the *done* state it will stay there and continually load the final value of *x* into the *gcd* output register by setting *gld* to 1.

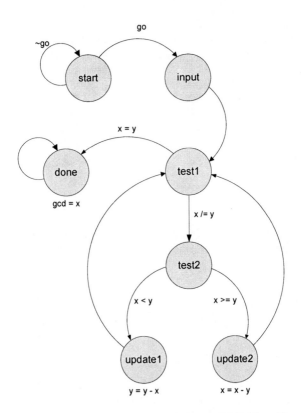

Figure 2.5 State diagram for the GCD algorithm

Listing 2.1 shows a VHDL program called *gcd_datapath.vhd* that implements the datapath in Fig. 2.3. This program uses the multiplexer component *mux2g* from Example 1 and the register component *reg* from Example 2. The two subtractors and two comparators are simply implemented within the program. Listing 2.2 shows a VHDL program called *gcd_control.vhd* for the control unit in Fig. 2.4, which implements the state diagram in Fig. 2.5 as a Moore machine. Listing 2.3 is a VHDL program called *gcd.vhd*, which simply wires together the datapath from Listing 2.1 with the control unit from Listing 2.2 to create the complete *gcd* algorithm described by Fig. 2.4.

Listing 2.1 gcd_datapath.vhd

```vhdl
-- Example 23a: gcd_datapath
library IEEE;
use IEEE.STD_LOGIC_1164.all;
use IEEE.STD_LOGIC_unsigned.all;

entity gcd_datapath is
  port(
        clk   : in STD_LOGIC;
        clr   : in STD_LOGIC;
        xmsel : in STD_LOGIC;
        ymsel : in STD_LOGIC;
        xld   : in STD_LOGIC;
        yld   : in STD_LOGIC;
        gld   : in STD_LOGIC;
        xin   : in STD_LOGIC_VECTOR(3 downto 0);
        yin   : in STD_LOGIC_VECTOR(3 downto 0);
        gcd   : out STD_LOGIC_VECTOR(3 downto 0);
        eqflg : out STD_LOGIC;
        ltflg : out STD_LOGIC
      );
end gcd_datapath;

architecture gcd_datapath of gcd_datapath is
  component mux2g
  generic(
        N : INTEGER);
  port(
        a : in std_logic_vector(N-1 downto 0);
        b : in std_logic_vector(N-1 downto 0);
        s : in std_logic;
        y : out std_logic_vector(N-1 downto 0));
  end component;

  component reg
  generic(
        N : INTEGER := 8);
  port(
        load : in std_logic;
        d    : in std_logic_vector(N-1 downto 0);
        clk  : in std_logic;
        clr  : in std_logic;
        q    : out std_logic_vector(N-1 downto 0));
  end component;

signal x, y, x1, y1, xmy, ymx: STD_LOGIC_VECTOR(3 downto 0);
begin
  xmy <= x - y;
  ymx <= y - x;
```

Listing 2.1 (cont.) gcd_datapath.vhd

```vhdl
  EQ: process(x,y)
  begin
        if x = y then
                eqflg <= '1';
        else
                eqflg <= '0';
        end if;
  end process EQ;

  LT: process(x,y)
  begin
        if x < y then
                ltflg <= '1';
        else
                ltflg <= '0';
        end if;
  end process LT;

M1 : mux2g
  generic map(
        N => 4
  )
  port map(
        a => xmy,
        b => xin,
        s => xmsel,
        y => x1
  );

M2 : mux2g
  generic map(
        N => 4
  )
  port map(
        a => ymx,
        b => yin,
        s => ymsel,
        y => y1
  );

R1 : reg
  generic map(
        N => 4
  )
  port map(
        load => xld,
        d => x1,
        clk => clk,
        clr => clr,
        q => x
  );
```

Listing 2.1(cont.) gcd_datapath.vhd

```vhdl
R2 : reg
  generic map(
        N => 4
  )
  port map(
        load => yld,
        d => y1,
        clk => clk,
        clr => clr,
        q => y
  );

R3 : reg
  generic map(
        N => 4
  )
  port map(
        load => gld,
        d => x,
        clk => clk,
        clr => clr,
        q => gcd
  );

end gcd_datapath;
```

Listing 2.2 gcd_control.vhd

```vhdl
-- Example 23b: gcd_control
library IEEE;
use IEEE.STD_LOGIC_1164.all;

entity gcd_control is
  port(
        clk : in STD_LOGIC;
        clr : in STD_LOGIC;
        go : in STD_LOGIC;
        eqflg : in STD_LOGIC;
        ltflg : in STD_LOGIC;
        xmsel : out STD_LOGIC;
        ymsel : out STD_LOGIC;
        xld : out STD_LOGIC;
        yld : out STD_LOGIC;
        gld : out STD_LOGIC
      );
end gcd_control;
```

Listing 2.2 (cont.) gcd_control.vhd

```vhdl
architecture gcd_control of gcd_control is
type state_type is (start, input, test1, test2,
                    update1, update2, done);
signal present_state, next_state: state_type;
begin

sreg: process(clk, clr)
begin
  if clr = '1' then
       present_state <= start;
  elsif clk'event and clk = '1' then
       present_state <= next_state;
  end if;
end process;

C1: process(present_state, go, eqflg, ltflg)
begin
   case present_state is
      when start =>
        if go = '1' then
             next_state <= input;
        else
             next_state <= start;
        end if;
      when input =>
        next_state <= test1;
      when test1 =>
        if eqflg = '1' then
             next_state <= done;
        else
             next_state <= test2;
        end if;
      when test2 =>
        if ltflg = '1' then
             next_state <= update1;
        else
             next_state <= update2;
        end if;
      when update1 =>
        next_state <= test1;
      when update2 =>
        next_state <= test1;
      when done =>
        next_state <= done;
      when others =>
        null;
   end case;
end process;
```

Listing 2.2 (cont.) gcd_control.vhd

```vhdl
C2: process(present_state)
begin
   xld <= '0'; yld <= '0'; gld <= '0';
   xmsel <= '0'; ymsel <= '0';
   case present_state is
       when input =>
           xld <= '1'; yld <= '1';
           xmsel <= '1'; ymsel <= '1';
       when update1 =>
           yld <= '1';
       when update2 =>
           xld <= '1';
       when done =>
           gld <= '1';
       when others =>
           null;
   end case;
end process;

end gcd_control;
```

Listing 2.3 gcd.vhd

```vhdl
-- Example 23c: gcd
library IEEE;
use IEEE.STD_LOGIC_1164.all;

entity gcd is
   port(
        clk : in STD_LOGIC;
        clr : in STD_LOGIC;
        go : in STD_LOGIC;
        xin : in STD_LOGIC_VECTOR(3 downto 0);
        yin : in STD_LOGIC_VECTOR(3 downto 0);
        gcd_out : out STD_LOGIC_VECTOR(3 downto 0)
       );
end gcd;

architecture gcd of gcd is
  component gcd_datapath
  port(
        clk : in std_logic;
        clr : in std_logic;
        xmsel : in std_logic;
        ymsel : in std_logic;
        xld : in std_logic;
        yld : in std_logic;
        gld : in std_logic;
        xin : in std_logic_vector(3 downto 0);
        yin : in std_logic_vector(3 downto 0);
        gcd : out std_logic_vector(3 downto 0);
        eqflg : out std_logic;
        ltflg : out std_logic);
  end component;
```

Listing 2.3 (cont.) gcd.vhd

```vhdl
component gcd_control
  port(
        clk : in std_logic;
        clr : in std_logic;
        go : in std_logic;
        eqflg : in std_logic;
        ltflg : in std_logic;
        xmsel : out std_logic;
        ymsel : out std_logic;
        xld : out std_logic;
        yld : out std_logic;
        gld : out std_logic);
  end component;

  signal eqflg, ltflg, xmsel, ymsel: std_logic;
  signal xld, yld, gld: std_logic;
begin

U1 : gcd_datapath
  port map(
        clk => clk,
        clr => clr,
        xmsel => xmsel,
        ymsel => ymsel,
        xld => xld,
        yld => yld,
        gld => gld,
        xin => xin,
        yin => yin,
        gcd => gcd_out,
        eqflg => eqflg,
        ltflg => ltflg
  );

U2 : gcd_control
  port map(
        clk => clk,
        clr => clr,
        go => go,
        eqflg => eqflg,
        ltflg => ltflg,
        xmsel => xmsel,
        ymsel => ymsel,
        xld => xld,
        yld => yld,
        gld => gld
  );

end gcd;
```

A simulation of the gcd algorithm given in Listing 2.3 is shown in Fig. 2.6. Note that this simulation shows the calculation of the GCD of 12 (hex C) and 8. The clock frequency is 50 MHz and the result 4 is obtained at 210 ns. Note in this simulation the values of *x* and *y* from the datapath and the signals *present_state* and *next_state* from the control unit are displayed. You can display any signal from any of the sub-modules in your program in you simulations. This makes it easy to debug your programs using the Aldec Active-HDL simulator.

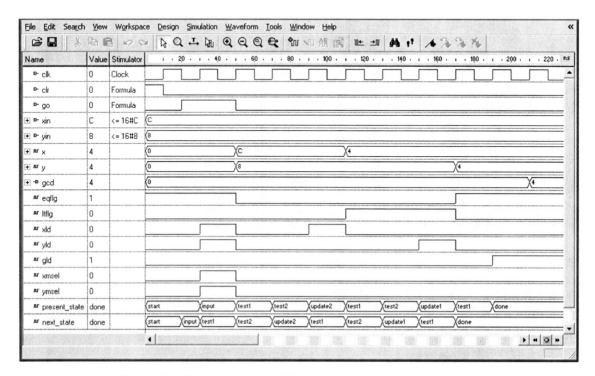

Figure 2.6 Simulation of the GCD algorithm in Listing 2.3

To test the program on the Nexys-2 board we will use the top-level design shown in Fig. 2.7. We will use the *clkdiv* component from Example 6 to generate a 25 MHz clock signal (*clk25*) and a 190 Hz clock signal (*clk190*). We will also use the *x7segb* component from Example 21 to display the answer on the right-most 7-segment display. The values of *xin*(3:0) and *yin*(3:0) will come from switches *sw*(7:4) and *sw*(3:0) respectively. These input values will be displayed on the eight LEDs. We will use *btn*(0) for the *go* signal and *btn*(3) for the *clr* signal.

Listing 2.4 is a VHDL program for this top-level design. Note that the output of the *gcd* component is called *gcd_out*(3:0) – we could not call it *gcd* because this is the name of the component – and this is connected to the signal *gcds*(3:0). This signal is then concatenated with 12 leading zeros to form the 16-bit input to the component *x7segb*.

Run this program and test it by changing the switch settings and pressing *btn*(0). Because you end up in the *done* state forever, you will need to press *btn*(3) to reset the program in order to run it again.

GCD Algorithm – Part 2

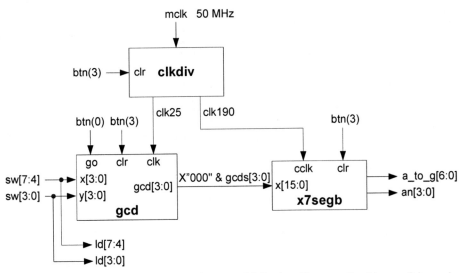

Figure 2.7 Top-level design for testing the GCD algorithm on the Nexys-2 board

Listing 2.4 gcd_top.vhd

```vhdl
-- Example 23d: gcd_top
library IEEE;
use IEEE.STD_LOGIC_1164.all;

entity gcd_top is
   port(
         mclk : in STD_LOGIC;
         btn : in STD_LOGIC_VECTOR(3 downto 0);
         sw : in STD_LOGIC_VECTOR(7 downto 0);
         ld : out STD_LOGIC_VECTOR(7 downto 0);
         a_to_g : out STD_LOGIC_VECTOR(6 downto 0);
         an : out STD_LOGIC_VECTOR(3 downto 0)
      );
end gcd_top;

architecture gcd_top of gcd_top is
  component clkdiv
  port(
        mclk : in std_logic;
        clr : in std_logic;
        clk25 : out std_logic;
        clk190 : out std_logic);
  end component;

  component gcd
  port(
        clk : in std_logic;
        clr : in std_logic;
        go : in std_logic;
        xin : in std_logic_vector(3 downto 0);
        yin : in std_logic_vector(3 downto 0);
        gcd_out : out std_logic_vector(3 downto 0));
  end component;
```

Listing 2.4 (cont.) gcd_top.vhd

```vhdl
component x7segb
  port(
        x : in std_logic_vector(15 downto 0);
        cclk : in std_logic;
        clr : in std_logic;
        a_to_g : out std_logic_vector(6 downto 0);
        an : out std_logic_vector(3 downto 0));
  end component;

signal clk25, clk190, clr: std_logic;
signal x: std_logic_vector(15 downto 0);
signal gcds: std_logic_vector(3 downto 0);
begin

clr <= btn(3);
x <= X"000" & gcds;
ld <= sw;

U1 : clkdiv
  port map(
        mclk => mclk,
        clr => clr,
        clk25 => clk25,
        clk190 => clk190
  );

U2 : gcd
  port map(
        clk => clk25,
        clr => clr,
        go => btn(0),
        xin => sw(7 downto 4),
        yin => sw(3 downto 0),
        gcd_out => gcds
  );

U3 : x7segb
  port map(
        x => x,
        cclk => clk190,
        clr => clr,
        a_to_g => a_to_g,
        an => an
  );

end gcd_top;
```

Example 24

An Integer Square Root Algorithm

The C algorithm shown in Fig. 2.8 performs an integer square root of the input *a* as shown in Table 2.1. Note from Table 2.1 that the difference between successive squares, *delta*, is just the sequence of odd numbers. Also note that the *while* loop is executed as long as *square* is less than or *equal* to *a*. Thus, the value of the square root, *delta*/2 – 1, occurs in the row following the *square* value in Table 2.1.

```c
unsigned long sqrt(unsigned long a){
   unsigned long square = 1;
   unsigned long delta = 3;
   while(square <= a){
      square = square + delta;
      delta = delta + 2;
   }
      return (delta/2 - 1);
}
```

Figure 2.8 Integer square root algorithm

Table 2.1
Illustrating the algorithm in Fig. 2.8

n	square = n^2	delta	delta/2-1
0	0		
1	1	3	
2	4	5	1
3	9	7	2
4	16	9	3
5	25	11	4
6	36	13	5
7	49	15	6
8	64	17	7
9	81	19	8
10	100	21	9
11	121	23	10
12	144	25	11
13	169	27	12
14	196	29	13
15	225	31	14
16	256	33	15
17	289		

Our goal is to implement this algorithm in hardware using the same method that we used in Example 23. The datapath for this square root algorithm is shown in Fig. 2.9. We will limit the input value *a*(7:0) to eight bits so we can connect them to the eight switches *sw*(7:0). Thus, the register *aReg* will be an 8-bit register. From Table 2.1 the maximum value of *delta* will be 33. This suggests that the size of the *delta* register should be 6 bits. However, when we are done the final value of *delta* is divided by 2 and then 1 is subtracted from this result. If the value is 33 but we only use a 5-bit register for *delta* then the value in *delta* will be 1. When we divide this by 2 we will get 0, and when we subtract 1 from 0 we get 11111, the lower four bits of which are equal to 15 – the correct result. Thus, we can get by with a 5-bit register for *delta*. However, from Table 2.1 the maximum value of *square* in the algorithm will be 256, and therefore the *square* register must be 9 bits wide. The maximum square root value will be 15 and thus the output will be stored in a 4-bit register.

In addition to the four registers *aReg*, *sqReg*, *delReg*, and *outReg*, this datapath contains four combinational components. The output, *lteflg*, of the <= component will be 1 if *square* (the output of *sqReg*) is less than or equal to *a* (the output of *aReg*).

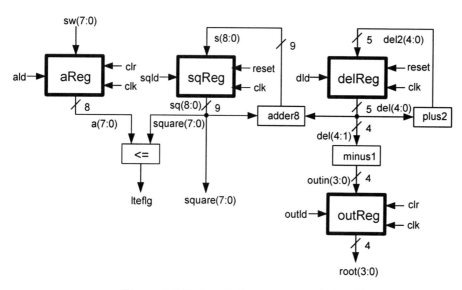

Figure 2.9 Datapath for square root algorithm

The VHDL program in Listing 2.5 implements the complete datapath shown in Fig. 2.9. The four registers have different bus widths and are initialized to different values. The generic register described in the VHDL file *regr2.vhd* shown in Listing 2.6 is used to instantiate all four registers in Listing 2.5. The two parameters *BIT0* and *BIT1* are used to set the lower two bits of the generic register on reset. Note in Listing 2.5 that the value of *sqReg* is initialized to 1 and the value of *delReg* is initialized to 3. The combinational components *adder8*, *plus2*, *minus1*, and *lte* are implemented by separate processes in Listing 2.5.

The output signal *lteflg* in Fig. 2.9 is sent to the control unit. The control unit will then provide the datapath with the load signals for all of the registers. The design of the control unit will be described next.

Listing 2.5 SQRTpath.vhd

```vhdl
-- Example 24a: Square root datapath
library IEEE;
use IEEE.STD_LOGIC_1164.ALL;
use IEEE.STD_LOGIC_ARITH.ALL;
use IEEE.STD_LOGIC_UNSIGNED.ALL;

entity SQRTpath is
    port ( clk : in std_logic;
           reset : in std_logic;
           ald : in std_logic;
           sqld : in std_logic;
           dld : in std_logic;
           outld : in std_logic;
           sw : in std_logic_vector(7 downto 0);
           lteflg : out std_logic;
           root : out std_logic_vector(3 downto 0)
    );
end SQRTpath;

architecture SQRTpath of SQRTpath is

      component regr2
      generic(N: positive;
                bit0: std_logic;
                bit1: std_logic);
      port (
              d: in STD_LOGIC_VECTOR (N-1 downto 0);
              load: in STD_LOGIC;
              reset: in STD_LOGIC;
              clk: in STD_LOGIC;
              q: out STD_LOGIC_VECTOR (N-1 downto 0)
      );
      end component;

      signal a: STD_LOGIC_VECTOR (7 downto 0);
      signal sq, s: STD_LOGIC_VECTOR (8 downto 0);
      signal del, dp2: STD_LOGIC_VECTOR (4 downto 0);
      signal dm1: STD_LOGIC_VECTOR (3 downto 0);
      constant bus_width9: integer := 9;
      constant bus_width8: integer := 8;
      constant bus_width5: integer := 5;
      constant bus_width4: integer := 4;

begin

    adder8: process(sq, del)
                begin
                    s <= sq + ("0000" & del);
                end process;

    plus2: process(del)
               begin
                   dp2 <= del + 2;
               end process;
```

Listing 2.5 (cont.) SQRTpath.vhd

```vhdl
    minus1: process(del)
                begin
                    dm1 <= del(4 downto 1) - 1;
                end process;

    lte: process(sq, a)
                begin
                    if(sq <= ('0' & a)) then
                        lteflg <= '1';
                    else
                        lteflg <= '0';
                    end if;
                end process;

  aReg: regr2 generic map(N => bus_width8, BIT0 => '0', BIT1 => '0')
  port map
  (d => sw, load =>ald,   reset => reset, clk =>clk, q => a);

  sqReg: regr2 generic map(N => bus_width9, BIT0 => '1', BIT1 => '0')
  port map
  (d => s, load => sqld,   reset => reset, clk =>clk, q => sq);

  delReg: regr2 generic map(N => bus_width5, BIT0 => '1', BIT1 => '1')
  port map
  (d => dp2, load => dld,   reset => reset, clk =>clk, q => del);

  outReg: regr2 generic map(N => bus_width4, BIT0 => '0', BIT1 => '0')
  port map
  (d => dm1, load => outld,   reset => reset, clk =>clk, q => root);

end SQRTpath;
```

Listing 2.6 regr2.vhd

```vhdl
--    Example 24b: N-bit register with reset and load
--    Resets to initial value of lowest 2 bits
library IEEE;
use IEEE.std_logic_1164.all;

entity regr2 is
    generic(N: integer;
            BIT0: std_logic;
            BIT1: std_logic);
    port (
        d: in STD_LOGIC_VECTOR (N-1 downto 0);
        load: in STD_LOGIC;
        reset: in STD_LOGIC;
        clk: in STD_LOGIC;
        q: out STD_LOGIC_VECTOR (N-1 downto 0)
    );
end regr2;
```

Listing 2.6 (cont.) regr2.vhd

```
architecture regr2 of regr2 is
begin
  process(clk, reset)
  begin
    if reset = '1' then
            q <= (others => '0');
            q(0) <= BIT0;
            q(1) <= BIT1;
    elsif (clk'event and clk = '1') then
            if load = '1' then
            q <= d;
            end if;
    end if;
  end process;
end regr2;
```

Square Root Control Unit

Listing 2.5 describes the datapath shown in Fig. 2.9. In this section we will design the control unit that will control this datapath as shown in Fig. 2.10. Note in this figure that the datapath sends the signal *lteflg* to the control unit, and the control unit sends the register load signals *ald*, *sqld*, *dld*, and *outld* to the datapath.

The control unit will be a *state machine* that defines the timing sequence of the algorithm. For this square root algorithm there are four states: *start*, *test*, *update*, and *done* as shown in Fig. 2.11. The program begins in the *start* state, and stays in this state until the *go* signal goes high. It then goes to the *test* state, which will test to see if *square* <= *a* is true or false. If it is true, then the program goes to the *update* state, which will update the values of *square* and *delta* in the *while* loop of Fig. 2.8. Otherwise, the program goes to the *done* state, which computes the return value and stays in this *done* state. Listing 2.7 is a VHDL program for the control unit in Fig. 2.10, which implements the state diagram shown in Fig. 2.11.

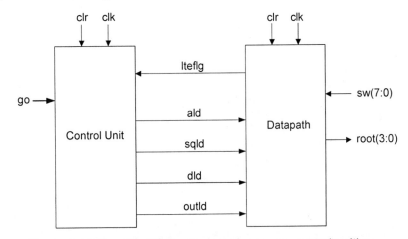

Figure 2.10 Top-level logic diagram for square root algorithm

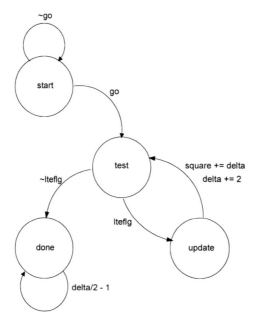

Figure 2.11 State diagram for square root algorithm

Note that we have implemented the state machine in Listing 2.7 as a Moore machine of the type shown in Fig. 1.37 with the three processes: the sequential state register block and the two combinational blocks *C1* and *C2*. Note how the *C1* process finds the next state by directly implementing the state diagram in Fig. 2.11 with a *case* statement. The output block *C2* also uses a *case* statement to set the register load signals to the proper values for each state. Verify that these load signals are appropriate for implementing the square root algorithm.

Listing 2.7 SQRTctrl.vhd

```
-- Example 24c: Square root control
library IEEE;
use IEEE.STD_LOGIC_1164.ALL;
use IEEE.STD_LOGIC_ARITH.ALL;
use IEEE.STD_LOGIC_UNSIGNED.ALL;

entity SQRTctrl is
    port ( clk   : in  std_logic;
           clr   : in  std_logic;
           lteflg: in  std_logic;
           go    : in  std_logic;
           ald   : out std_logic;
           sqld  : out std_logic;
           dld   : out std_logic;
           outld : out std_logic
         );
end SQRTctrl;

architecture SQRTctrl of SQRTctrl is
```

Listing 2.7 (cont.) SQRTctrl.vhd

```vhdl
type state_type is (start, test, update, done);
signal present_state, next_state: state_type;
begin
sreg: process(clk, clr)
begin
  if clr = '1' then
      present_state <= start;
  elsif clk'event and clk = '1' then
      present_state <= next_state;
  end if;
end process;

C1: process(present_state, go, lteflg)
begin
   case present_state is
       when start =>
              if go = '1' then
                    next_state <= test;
              else
                    next_state <= start;
              end if;
       when test =>
              if lteflg = '1' then
                    next_state <= update;
              else
                    next_state <= done;
              end if;
       when update =>
              next_state <= test;
       when done =>
              next_state <= done;
       when others =>
              null;
       end case;
end process;


C2: process(present_state)
begin
   ald <= '0'; sqld <= '0'; dld <= '0'; outld <= '0';
   case present_state is
             when start =>
                    ald <= '1';
             when test =>
                    null;
             when update =>
                    sqld <= '1'; dld <= '1';
             when done =>
                    outld <= '1';
             when others =>
                    null;
   end case;
end process;

end SQRTctrl;
```

The top-level design shown in Fig. 2.10 can be implemented in VHDL by simply instantiating the datapath and control unit as shown in Listing 2.8. Note that we have included an output signal *done* that goes high when the *done* state is entered. A simulation of this program that computes the square root of 64 is shown in Fig. 2.12.

Listing 2.8 SQRT.vhd

```vhdl
-- Example 24d: Integer Square Root
library IEEE;
use IEEE.STD_LOGIC_1164.ALL;
use IEEE.STD_LOGIC_ARITH.ALL;
use IEEE.STD_LOGIC_UNSIGNED.ALL;

entity sqrt is
    port ( clk : in std_logic;
           clr : in std_logic;
           go  : in std_logic;
           sw  : in std_logic_vector(7 downto 0);
           done : out std_logic;
           root : out std_logic_vector(3 downto 0)
           );
end sqrt;

architecture sqrt of sqrt is

    component SQRTctrl is
    port ( clk : in std_logic;
        clr : in std_logic;
        lteflg : in std_logic;
        strt : in std_logic;
        ald : out std_logic;
        sqld : out std_logic;
        dld : out std_logic;
        outld : out std_logic
           );
    end component;


    component SQRTpath
    port ( clk : in std_logic;
        reset : in std_logic;
        ald : in std_logic;
        sqld : in std_logic;
        dld : in std_logic;
        outld : in std_logic;
        S : in std_logic_vector(7 downto 0);
        lteflg : out std_logic;
        root : out std_logic_vector(3 downto 0)
           );
    end component;

    signal lteflg, ald, sqld, dld, outld: std_logic;
```

Listing 2.8 (cont.) SQRT.vhd

```
begin
done <= outld;

sqrt1: SQRTctrl port map
   (clk => clk, clr => clr, lteflg => lteflg, go => go, ald => ald,
      sqld => sqld, dld => dld, outld => outld);

sqrt2: SQRTpath port map
   ( clk => clk, reset => clr, ald => ald, sqld => sqld, dld => dld,
      outld => outld, S => sw, lteflg => lteflg, root => root);

end sqrt;
```

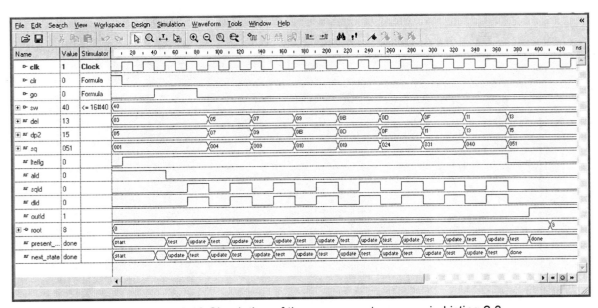

Figure 2.12 Simulation of the square root program in Listing 2.8

To test the square root algorithm on the Nexys-2 board we will use the top-level design shown in Fig. 2.13. When *btn*(3) is pressed to reset the circuit the decimal value of the eight switch settings is displayed on the 7-segment display. We use the 8-bit binary-to-BCD program in Listing 1.14 of Example 12 to do this conversion. When *btn*(0) is pressed the *sqrt* component computes the square root of the switch settings, and when the calculation is complete the *done* signal switches the multiplexer so that the decimal value of the square root is displayed on the 7-segment display. Try it.

86   Example 24

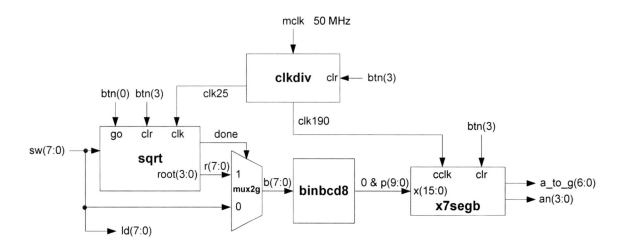

Figure 2.13 Top-level design for testing square root algorithm

Listing 2.9 sqrt_top.vhd

```vhdl
-- Example 24e: sqrt_top
library IEEE;
use IEEE.STD_LOGIC_1164.all;

entity sqrt_top is
        port(
            mclk : in STD_LOGIC;
            btn : in STD_LOGIC_VECTOR(3 downto 0);
            sw : in STD_LOGIC_VECTOR(7 downto 0);
            ld : out STD_LOGIC_VECTOR(7 downto 0);
            a_to_g : out STD_LOGIC_VECTOR(6 downto 0);
            an : out STD_LOGIC_VECTOR(3 downto 0)
        );
end sqrt_top;

architecture sqrt_top of sqrt_top is
    component clkdiv
    port(
        mclk : in std_logic;
        clr : in std_logic;
        clk25 : out std_logic;
        clk190 : out std_logic);
    end component;

    component sqrt
    port(
        clk : in std_logic;
        clr : in std_logic;
        go : in std_logic;
        done : out std_logic;
        sw : in std_logic_vector(7 downto 0);
        root : out std_logic_vector(3 downto 0));
    end component;
```

Listing 2.9 (cont.) sqrt_top.vhd

```vhdl
        component mux2g
        generic(
                N : INTEGER);
        port(
                a : in std_logic_vector(N-1 downto 0);
                b : in std_logic_vector(N-1 downto 0);
                s : in std_logic;
                y : out std_logic_vector(N-1 downto 0));
        end component;

        component binbcd8
        port(
                b : in std_logic_vector(7 downto 0);
                p : out std_logic_vector(9 downto 0));
        end component;

        component x7segb
        port(
                x : in std_logic_vector(15 downto 0);
                cclk : in std_logic;
                clr : in std_logic;
                a_to_g : out std_logic_vector(6 downto 0);
                an : out std_logic_vector(3 downto 0));
        end component;
signal clk25, clk190, clr, done: std_logic;
signal x: std_logic_vector(15 downto 0);
signal p: std_logic_vector(9 downto 0);
signal root: std_logic_vector(3 downto 0);
signal b, r: std_logic_vector(7 downto 0);
begin

clr <= btn(3);
r <= "0000" & root;
x <= "000000" & p;
ld <= sw;

U1 : clkdiv
     port map(
            mclk => mclk,
            clr => clr,
            clk25 => clk25,
            clk190 => clk190
     );

U2 : sqrt
     port map(
            clk => clk25,
            clr => clr,
            go => btn(0),
            done => done,
            sw => sw,
            root => root
     );
```

Listing 2.9 (cont.) sqrt_top.vhd

```vhdl
U3 : mux2g
    generic map(
        N => 8
    )
    port map(
        a => sw,
        b => r,
        s => done,
        y => b
    );

U4 : binbcd8
    port map(
        b => b,
        p => p
    );

U5 : x7segb
    port map(
        x => x,
        cclk => clk190,
        clr => clr,
        a_to_g => a_to_g,
        an => an
    );
end sqrt_top;
```

3

Integrating the Datapath and Control Unit

In Example 22 in Section 1.8 we saw that a *while* loop could not be synthesized directly in VHDL. In Chapter 2 we saw how we could use a state diagram to implement a *while* loop. However, the price of implementing the simple algorithm shown in Fig. 3.1 was to write a total of five *.vhd* files: *mux2g.vhd* in Listing 1.1, *reg.vhd* in Listing 1.3, *gcd_datapath.vhd* in Listing 2.1, *gcd_control.vhd* in Listing 2.2, and *gcd.vhd* in Listing 2.3. That seems like a lot of effort compared to the direct VHDL implementation given by the file *gcd1.vhd* in Listing 1.30 in Example 22, which simulated fine, but did not synthesize. Is there any way to modify the single file *gcd1.vhd* in Listing 1.30 so that it will synthesize?

The problem is that the process in Listing 1.30 is a combinational circuit that depends only on the values of *x* and *y*. But we saw in Example 23 that *x* and *y* needed to be registers that could store intermediate values as the algorithm executed. Instead of using separate components for the multiplexers and registers in the datapath of Fig. 2.3 we could infer these registers, as well as the subtractors and comparators by using a process as shown in Fig. 3.2.

```
while(~go){}
x = xin;
y = yin;
while (x /= y) {
    if(x < y)
        y = y - x;
    else
        x = x - y;
}
gcd = x;
while(1){}
```

Figure 3.1 Euclid's GCD algrorithm

```
A1: process(clk, clr)
begin
    if clr = '1' then
        << asynchronous clears >>
    elsif clk'event and clk = '1' then
        if x = y then
            gcd <= x;
        elsif x < y then
            y <= y - x;
        else
            x <= x - y;
        end if;
    end if;
end process A1;
```

Figure 3.2 Process for implementing a *while* loop

Recall that when we use the phrase *clk'event* **and** *clk* = '1' all the the signals or variables on the left-hand side of following assignment statements will be registered. We have also used an *if* statement (**if** *x* = *y*) instead of the *while* statement in Fig. 3.1. In general, you can implement a *while* loop with an *if* statement, but you would need some type of *goto* statement at the end of the loop to go back to the beginning of the loop. But note that in Fig. 3.2 all of the statements following the phrase *clk'event* **and** *clk* = '1' will be executed on every rising edge of the clock. Thus, the *while* test given by the *if* statement will continually be tested. But how do we stop testing it when the *while* loop is exited? Also, how do we know when to start the algorithm – i.e. how do we know when to leave the *start* state in the state diagram in Fig. 2.5?

In general, we will use a *go* signal to start an algorithm and a *done* signal to indicate when the algorithm is complete as shown in Fig. 3.3. The *go* signal needs to be a single pulse that is one clock cycle wide. Depending on the application the *done* signal can be a single pulse (that could be used as the *go* input to another algorithm) or it could just go high. Fig. 3.4 shows how a pulse *done* signal can be generated within a general algorithm process. Note that the *calc* variable goes high during the algorithm calculation. Example 26 will show an example if creating a *done* signal that stays high.

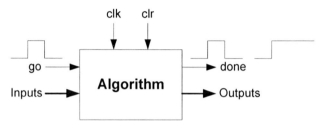

Figure 3.3 An algorithm with *go* and *done* signals

```
A2: process(clk, clr)
variable calc, donev: std_logic;
begin
    if clr = '1' then
        << asynchronous clears >>
        donev := '0';
        calc := '0';
    elsif clk'event and clk = '1' then
        donev := '0';
        if go = '1' then
            << initialize signals and variables >>
            calc := '1';
        elsif calc = '1' then
            if <<done condition>> then
                << assign outputs >>
                donev := '1';
                calc := '0';
            else
                << algorithm statements >>
            end if;
        end if;
    end if;
    done <= donev;
end process A2;
```

Figure 3.4 Generating the *done* output

Example 25

GCD Algorithm – Part 3

Listing 3.1 combines the code from Figs. 3.2 and 3.4 to form a single VHDL file, *gcd3.vhd*, that implements the GCD algorithm in Fig. 3.1. A simulation of this program is shown in Fig. 3.5. Note that this single file produces the same result as the five files in Listings 1.1, 1.3, 2.1, 2.2, and 2.3 used in Example 23 that produced the simulation shown in Fig. 2.6.

We can test the component *gcd3* given in Listing 3.1 on the Nexys-2 board by using the top-level design shown in Fig. 3.6. Note that the *go* input signal to the *gcd3* component must go high for only one clock cycle of the 25 MHz clock. When *go* is 1 then on the next rising edge of the clock the values of *x* and *y* get initialized to *xin* and *yin* and *calc* gets set to 1. Thus, on the next rising edge of the clock *go* will have gone to 0 so that the rest of the algorithm gets executed as long as *calc* is 1. Note that *calc* goes to 0 when *x* becomes equal to *y* and the *gcd* register gets the last value of *x*. At this point the variable *donev* gets set to 1, but will go to zero on the next rising edge of the clock because *calc* is now zero.

We will generate the *go* signal by using the *clock_pulse* component from Listing 1.5 of Example 4 with the 25 MHz clock. Inasmuch as the 25 MHz clock is controlling the *gcd3* component we need the *go* signal to be high for only one of these clock cycles. However, we still need to debounce *btn*(0) using the slower 190 Hz clock, and we do this with the debounce component from Listing 1.4 of Example 3. The VHDL program for the top-level design in Fig. 3.6 is given in Listing 3.2.

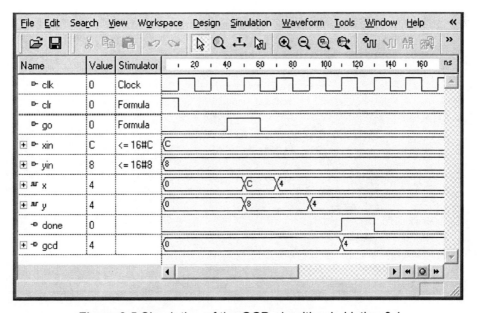

Figure 3.5 Simulation of the GCD algorithm in Listing 3.1

Listing 3.1 gcd3.vhd

```vhdl
-- Example 25a: gcd3
library IEEE;
use IEEE.STD_LOGIC_1164.all;
use IEEE.STD_LOGIC_unsigned.all;

entity gcd3 is
        port(
                clk : in STD_LOGIC;
                clr : in STD_LOGIC;
                go  : in STD_LOGIC;
                xin : in STD_LOGIC_VECTOR(3 downto 0);
                yin : in STD_LOGIC_VECTOR(3 downto 0);
                done : out STD_LOGIC;
                gcd : out STD_LOGIC_VECTOR(3 downto 0)
             );
end gcd3;

architecture gcd3 of gcd3 is
signal x, y: std_logic_vector(3 downto 0);
begin

G3: process(clk, clr)
variable calc, donev: std_logic;
begin
        if clr = '1' then
                x <= (others => '0');
                y <= (others => '0');
                gcd <= (others => '0');
                donev := '0';
                calc := '0';
        elsif clk'event and clk = '1' then
                donev := '0';
                if go = '1' then
                        x <= xin;
                        y <= yin;
                        calc := '1';
                elsif calc = '1' then
                        if x = y then
                                gcd <= x;
                                donev := '1';
                                calc := '0';
                        elsif x < y then
                                y <= y - x;
                        else
                                x <= x - y;
                        end if;
                end if;
        end if;
        done <= donev;
end process G3;

end gcd3;
```

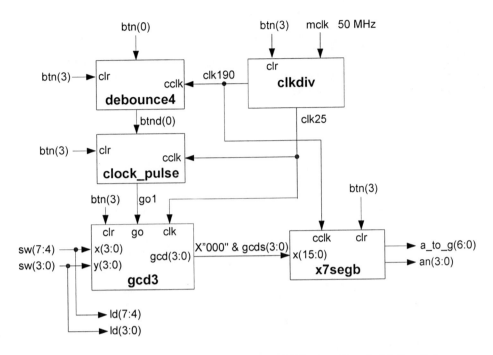

Figure 3.6 Top-level design for testing the GCD algorithm in Listing 3.1

Did the simpler approach in Listing 3.1 actually cost us anything? Table 3.1 compares the FPGA resources from the map and place and route reports used in both examples. In both cases we implemented only the *gcd* components and not the other components in Fig. 3.6. Note that Example 25 uses two less slices and five less total flip-flops. However, from the place and route reports we found that the maximum pin delay for Example 25 was 3.075 ns, which means that the algorithm should run at up to 325 MHz, while for Example 23 the maximum pin delay was 2.306, which means that that implementation should run at up to 433 MHz In general, the synthesis tools did a pretty good job of automatically converting our VHDL program in Listing 3.1 into the equivalent of the datapath and control unit that we designed in Example 23.

Table 3.1 Comparison of FPGA resource usage

	Example 23 gcd	Example 25 gcd3
Total number of slices	16	14
Total number if 4-input LUTs	28	28
Total number of slice flip-flops	15	8
Total number of IOB flip-flops	4	6
Total equivalent gate count	323	283
Maximum pin delay	2.306 ns	3.075 ns

Listing 3.2 gcd3_top.vhd

```vhdl
-- Example 25b: gcd3_top
library IEEE;
use IEEE.STD_LOGIC_1164.all;
use work.gcd3_components.all;

entity gcd3_top is
    port(
        mclk : in STD_LOGIC;
        btn : in STD_LOGIC_VECTOR(3 downto 0);
        sw : in STD_LOGIC_VECTOR(7 downto 0);
        ld : out STD_LOGIC_VECTOR(7 downto 0);
        a_to_g : out STD_LOGIC_VECTOR(6 downto 0);
        an : out STD_LOGIC_VECTOR(3 downto 0)
      );
end gcd3_top;

architecture gcd3_top of gcd3_top is

signal clk25, clk190, clr, go1, done: std_logic;
signal x: std_logic_vector(15 downto 0);
signal gcds, btnd: std_logic_vector(3 downto 0);
begin

clr <= btn(3);
x <= X"000" & gcds;
ld <= sw;

U1 : clkdiv
     port map(mclk => mclk, clr => clr, clk25 => clk25,
              clk190 => clk190);

U2 : debounce4
     port map(clk => clk190, clr => clr, inp => btn,
              outp => btnd);

U3 : clock_pulse
     port map(inp => btnd(0), cclk => clk25, clr => clr,
              outp => go1);

U4 : gcd3
     port map(
         clk => clk25, clr => clr, go => go1,
         xin => sw(7 downto 4), yin => sw(3 downto 0),
         done => done, gcd => gcds);

U5 : x7segb
     port map(x => x, cclk => clk190, clr => clr,
              a_to_g => a_to_g, an => an);

end gcd3_top;
```

Example 26

Integer Square Root – Part 2

In this example we will write a single VHDL program to implement the integer square root algorithm shown in Fig. 3.7 by using the method described at the beginning of this chapter and used for the GCD algorithm in Example 25. Listing 3.3 shows the program *sqrt2.vhd* that will do this. Compare this program with Listing 3.1 for the GCD algorithm. The main difference – apart from the different algorithm – is the fact that the *done* signal stays high after the square root is calculated. This is accomplished by simply changing the location of the statement
$$donev := '0';$$

A simulation of the program in Listing 3.3, which computes the square root of 64, is shown in Fig. 3.8.

```
unsigned long sqrt(unsigned long a){
  unsigned long square = 1;
  unsigned long delta = 3;
  while(square <= a){
      square = square + delta;
      delta = delta + 2;
  }
      return (delta/2 - 1);
}
```

Figure 3.7 Integer square root algorithm

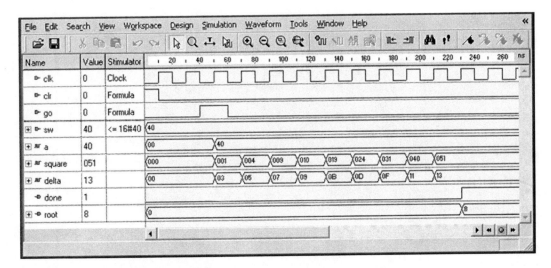

Figure 3.8 Simulation of the *sqrt2* program in Listing 3.3

Listing 3.3 sqrt2.vhd

```vhdl
-- Example 26a: sqrt2
library IEEE;
use IEEE.STD_LOGIC_1164.all;
use IEEE.STD_LOGIC_unsigned.all;

entity sqrt2 is
    port(
            clk  : in  STD_LOGIC;
            clr  : in  STD_LOGIC;
            go   : in  STD_LOGIC;
            sw   : in  STD_LOGIC_VECTOR(7 downto 0);
            done : out STD_LOGIC;
            root : out STD_LOGIC_VECTOR(3 downto 0)
        );
end sqrt2;

architecture sqrt2 of sqrt2 is
signal a: std_logic_vector(7 downto 0);
signal square: std_logic_vector(8 downto 0);
signal delta: std_logic_vector(4 downto 0);
begin

SR2: process(clk, clr)
variable calc, donev: std_logic;
begin
    if clr = '1' then
        a <= (others => '0');
        square <= (others => '0');
        delta <= (others => '0');
        root <= (others => '0');
        donev := '0';
        calc := '0';
    elsif clk'event and clk = '1' then
        if go = '1' then
            a <= sw;
            square <= "000000001";
            delta <= "00011";
            calc := '1';
            donev := '0';
        elsif calc = '1' then
            if square > a then
                root <= delta(4 downto 1) - 1;
                donev := '1';
                calc := '0';
            else
                square <= square + delta;
                delta <= delta + 2;
            end if;
        end if;
    end if;
    done <= donev;
end process SR2;

end sqrt2;
```

The top-level design for testing the *sqrt2* algorithm on the Nexys-2 board is shown in Fig. 3.9. Note that we use the *debounce4* and *clock_pulse* component as we did in Fig. 3.6 to generate a one-clock-pulse signal for the *go* input to the *sqrt2* component. We made the *done* signal stay high after the square root is calculated so that the multiplexer will display the square root value on the 7-segment display. You would press the reset *btn*(3) to display the switch values again, which you could change and then compute another square root by pressing *btn*(0). The VHDL program for the top-level design in Fig. 3.9 is given in Listing 3.4.

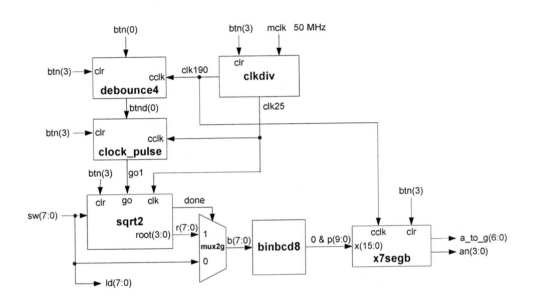

Figure 3.9 Top-level design for testing the *sqrt2* algorithm in Listing 3.3

Table 3.2 compares the FPGA resources used by this *sqrt2* implementation with the datapath–control unit implementation of Example 24. Note in this case that Example 26 uses slightly more slices and LUTs than Example 24, but can execute faster. We see from both Tables 3.1 and 3.2 that using more area in the FPGA (more slices and flip-flops) often leads to faster execution speed (shorter pin delays).

Table 3.2 Comparison of FPGA resource usage

	Example 24 sqrt	Example 26 sqrt2
Total number of slices	16	21
Total number if 4-input LUTs	29	39
Total number of slice flip-flops	15	14
Total number of IOB flip-flops	12	14
Total equivalent gate count	444	512
Maximum pin delay	2.178 ns	1.941 ns

Listing 3.4 sqrt2_top.vhd

```vhdl
-- Example 26b: sqrt2_top
library IEEE;
use IEEE.STD_LOGIC_1164.all;
use work.sqrt2_components.all;

entity sqrt2_top is
      port(
              mclk : in STD_LOGIC;
              btn : in STD_LOGIC_VECTOR(3 downto 0);
              sw : in STD_LOGIC_VECTOR(7 downto 0);
              ld : out STD_LOGIC_VECTOR(7 downto 0);
              a_to_g : out STD_LOGIC_VECTOR(6 downto 0);
              an : out STD_LOGIC_VECTOR(3 downto 0)
          );
end sqrt2_top;

architecture sqrt2_top of sqrt2_top is
signal clk25, clk190, clr, go1, done: std_logic;
signal x: std_logic_vector(15 downto 0);
signal p: std_logic_vector(9 downto 0);
signal root, btnd: std_logic_vector(3 downto 0);
signal b, r: std_logic_vector(7 downto 0);
begin

clr <= btn(3);
r <= "0000" & root;
x <= "000000" & p;
ld <= sw;

U1 : clkdiv
      port map(mclk => mclk, clr => clr, clk25 => clk25,
           clk190 => clk190);

U2 : debounce4
      port map(cclk => clk190, clr => clr, inp => btn,
           outp => btnd);

U3 : clock_pulse
      port map(inp => btnd(0), cclk => clk25, clr => clr,
           outp => go1);

U4 : sqrt2
      port map(clk => clk25, clr => clr, go => go1,
           done => done, sw => sw, root => root);

U5 : mux2g generic map(N => 8)
      port map(a => sw, b => r, s => done, y => b);

U6 : binbcd8
      port map(b => b, p => p);

U7 : x7segb
      port map(x => x, cclk => clk190, clr => clr,
           a_to_g => a_to_g, an => an);

end sqrt2_top;
```

4

Memory

In this chapter we will show how to implement memory on the Nexys-2 board by using the following six examples:

> Example 27 – A VHDL ROM
> Example 28 – Distributed RAM/ROM
> Example 29 – A Stack
> Example 30 – Block RAM
> Example 31 – External RAM
> Example 32 – External Flash Memory

Example 27 will show how to implement a read-only memory (ROM) using VHDL. This will be useful for small ROMs. For larger memories you can use the Core Generator to implement either distributed RAM (or ROM) that uses the FPGA LUTs (Example 28) or Block RAM that uses the block RAM within the FPGA (Example 30). For even larger memories you can access the 16 Mbytes of external RAM (Example 31) or 16 Mbytes of external flash memory (Example 32), which are on the Nexys-2 board.

Example 29 shows how to use distributed RAM to implement a stack. This stack will play a central role in the Forth core described in Chapter 9. You can skip this example for now if you wish.

Example 27

A VHDL ROM

In this example we will write a VHDL program to implement a read-only memory (ROM). A ROM can be a combinational component in which the output depends only on the input. For example, a ROM that contains eight bytes is shown in Fig. 4.1. The input is a 3-bit address, *addr*(2:0) and the output *M*(7:0) is the 8-bit contents of the ROM at the address *addr*(2:0). For example, if *addr*(2:0) = "110" then the output *M*(7:0) will be X"6C" or the contents of address 6.

Listing 4.1 shows how we can implement this ROM in VHDL. We first define the contents on the ROM as the eight constants *data*0 – *data*7. Note that we can define them as either binary or hex values. We then define a new data type called *rom_array*, which is an array of unspecified length (*NATURAL range* <>) in which each element of the array is of type *STD_LOGIC_VECTOR* (7 *downto* 0). We next create an instance of this *rom_array* by defining a constant called *rom* that is of type *rom_array*. The values of the elements in this *rom_array* are the constants *data*0 – *data*7.

The architecture then contains a single process that depends on the input *addr*. The integer *j* is set equal to the integer equivalent of the 3-bit value of *addr* and the output *M* is just equal to *rom(j)*. A simulation of this ROM program is shown in Fig. 4.2.

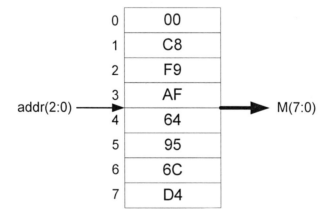

Figure 4.1 A ROM containing eight bytes

Listing 4.1 rom8.vhd

```vhdl
-- Example 27: ROM
library IEEE;
use IEEE.std_logic_1164.all;
use IEEE.std_logic_unsigned.all;

entity rom8 is
    port (
        addr: in STD_LOGIC_VECTOR (2 downto 0);
        M: out STD_LOGIC_VECTOR (7 downto 0)
    );
end rom8;

architecture rom8 of rom8 is
constant data0: STD_LOGIC_VECTOR (7 downto 0) := "00000000";
constant data1: STD_LOGIC_VECTOR (7 downto 0) := "11001000";
constant data2: STD_LOGIC_VECTOR (7 downto 0) := X"F9";
constant data3: STD_LOGIC_VECTOR (7 downto 0) := X"AF";
constant data4: STD_LOGIC_VECTOR (7 downto 0) := X"64";
constant data5: STD_LOGIC_VECTOR (7 downto 0) := X"95";
constant data6: STD_LOGIC_VECTOR (7 downto 0) := "01101100";
constant data7: STD_LOGIC_VECTOR (7 downto 0) := "11010100";

type rom_array is array (NATURAL range <>) of
                                STD_LOGIC_VECTOR (7 downto 0);
constant rom: rom_array := (
  data0, data1, data2, data3,
  data4, data5, data6, data7
  );

begin
  process(addr)
  variable j: integer;
  begin
      j := conv_integer(addr);
      M <= rom(j);
  end process;
end rom8;
```

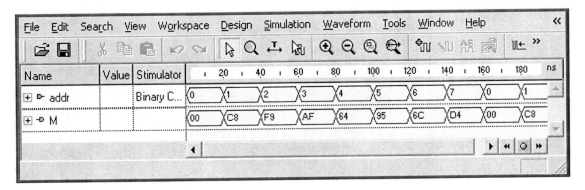

Figure 4.2 Simulation of the ROM program in Listing 4.1

To test the *rom8.vhd* component given in Listing 4.1 we add it to the *gcd3_top* program from Example 25. Instead of using the switches to set the two 4-bit values of *x* and *y* for the *gcd3* component we will use the 8-bit values stored in the ROM in Fig. 4.1. As we cycle through all eight addresses we will compute the greatest common divisor of the upper 4 bits and lower 4 bits of each byte in the ROM. We will get the address for the ROM from a 3-bit counter as shown in Fig. 4.3. For this counter we will use the *N*-bit *counter.vhd* component from Listing 1.6 in Example 5. We will use the *go1* signal in Fig. 4.3 as the clock input to the counter. This is a single clock pulse that gets generated each time you press *btn*(0). This signal will not only advance the address of the ROM by 1 but will start the calculation of the GCD of the ROM output for that new address.

On reset when you press *btn*(3) the output of the counter will be zero and the contents of the ROM at address 0 is also zero so all LEDs will be off. When you press *btn*(0) the address from the counter will go to 1 and the output *M* will be C8 and displayed on the LEDs. The output of the GCD calculation will be 4 and will be displayed on the 7-segment display. As you continue to press *btn*(0) you will cycle through all eight values in the ROM, each time displaying the GCD value on the 7-segment display.

The VHDL program for the top-level design shown in Fig. 4.3 is given in Listing 4.2.

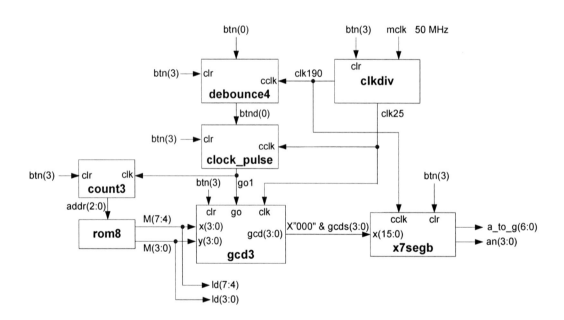

Figure 4.3 Top-level design for testing the ROM program in Listing 4.1

Listing 4.2 rom8_top.vhd

```vhdl
-- Example 27b: rom_top
library IEEE;
use IEEE.STD_LOGIC_1164.all;
use work.rom8_components.all;

entity rom8_top is
    port(
            mclk : in STD_LOGIC;
            btn : in STD_LOGIC_VECTOR(3 downto 0);
            ld : out STD_LOGIC_VECTOR(7 downto 0);
            a_to_g : out STD_LOGIC_VECTOR(6 downto 0);
            an : out STD_LOGIC_VECTOR(3 downto 0)
         );
end rom8_top;

architecture rom8_top of rom8_top is

signal clk25, clk190, clr, go1, done: std_logic;
signal x: std_logic_vector(15 downto 0);
signal addr: std_logic_vector(2 downto 0);
signal M: std_logic_vector(7 downto 0);
signal gcds, btnd: std_logic_vector(3 downto 0);
begin

clr <= btn(3);
x <= X"000" & gcds;
ld <= M;

U1 : clkdiv
     port map(mclk => mclk, clr => clr, clk25 => clk25,
              clk190 => clk190);

U2 : debounce4
     port map(clk => clk190, clr => clr, inp => btn,
              outp => btnd);

U3 : clock_pulse
     port map(inp => btnd(0), cclk => clk25, clr => clr,
              outp => go1);

U4 : gcd3
     port map(
              clk => clk25, clr => clr, go => go1,
              xin => M(7 downto 4), yin => M(3 downto 0),
              done => done, gcd => gcds);

U5 : x7segb
     port map(x => x, cclk => clk190, clr => clr,
              a_to_g => a_to_g, an => an);

U6 : counter generic map(N => 3)
     port map(clr => clr, clk => go1, q => addr);

U7 : rom
     port map(addr => addr, M => M);

end rom8_top;
```

Example 28

Distributed RAM/ROM

In this example we will show how to use the Core Generator to create a 16x8 distributed ROM. Distributed RAMs and ROMs use the LUTs in the FPGA to implement memory. To start the Core Generator select *design flow – Tools –* and then *CoreGen & Architecture Wizard*.

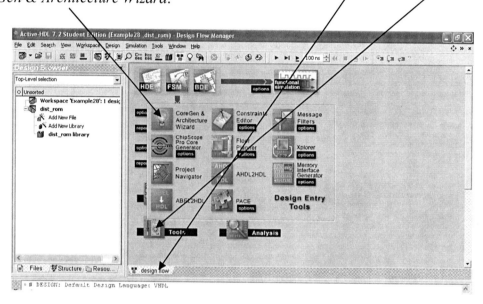

Click *Run CORE Generator*

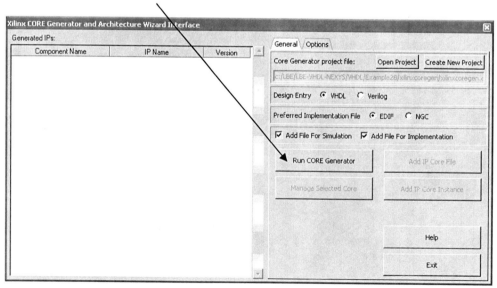

Distributed RAM/ROM

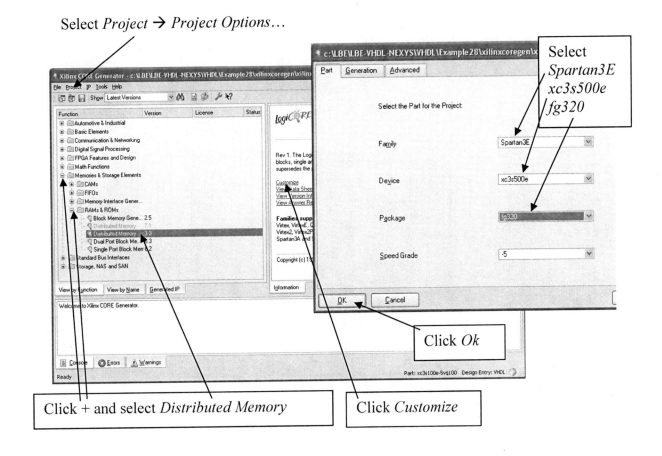

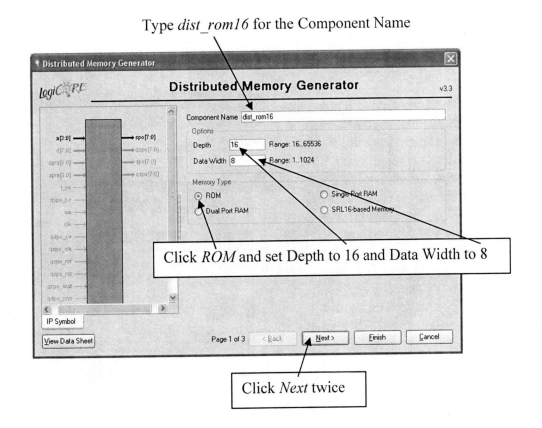

The contents of the ROM can be defined using a *.coe* file with the format shown in Listing 4.3. Click *Browse* to find this file and then click *Finish*.

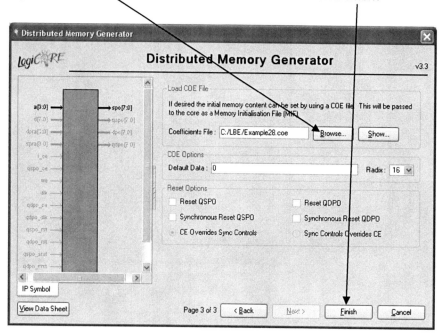

Listing 4.3 Example28.coe
```
; Example 28 Initialization file for a 16x8 distributed ROM
memory_initialization_radix = 16;
memory_initialization_vector =
0 C8 F9 AF
64 95 6C D4
39 E7 5A 96
84 37 28 4C;
```

The component *dist_rom16* will be created and all the files shown in the *Readme File* will be generated. These files will be in the *xilinxcoregen* folder in your project. You must add the file *dist_rom16.vhd* to your project. This file is only used for simulation as shown in Fig. 4.4. To synthesize the component *dist_rom16* you must copy the two files *dist_rom16.edn* and *dist_rom16.mif* from the *xilinxcoregen* folder to the *src* folder in your project. The file *dist_rom16.vho* contains the component declaration for *dist_rom* and a template for the component instantiation.

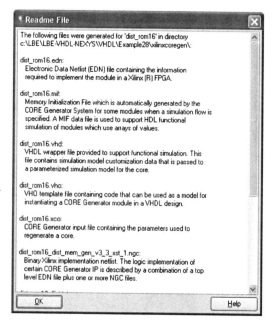

After generating the component *dist_rom16* click *Exit*.

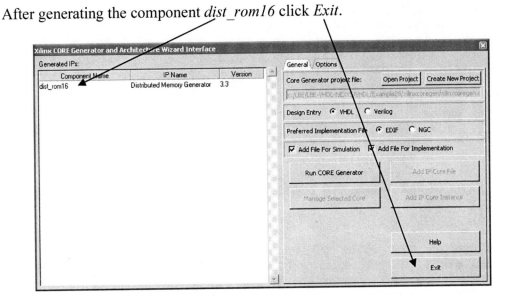

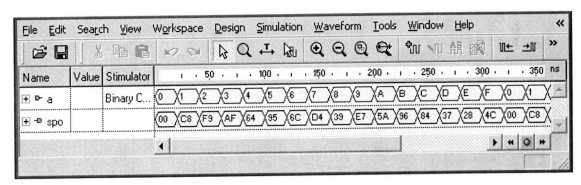

Figure 4.4 Simulation of the file *dist_rom16.vhd* generated by the Core Generator

You can test this *dist_rom16* component on the Nexys-2 board by replacing the *rom8* component in Fig. 4.3 with the *dist_rom16* component. The only other change is to make the counter a 4-bit counter with an output *addr*(3:0). We needed to make the depth of the *dist_rom16* component 16 bits because this is the minimum depth allowed in the Core Generator. The component declaration for *dist_rom16* is added to the *rom_components* package and the instantiation of the *dist_rom16* component is shown as U7 in the top-level VHDL program given in Listing 4.4. When you run this program you can step through all 16 values in the ROM calculating the GCD in each case.

The NEXYS2 board from Digilent contains a Xilinx Spartan3E-500 FG320 FPGA. This chip contains 1,164 CLBs arranged as 46 rows and 34 columns. There are therefore 4,656 slices with a total of 9,312 LUTs and flip-flops. Half of the LUTs on the chip can be used for a maximum of 74,752 bits of distributed RAM/ROM. Thus the largest distributed RAM/ROM you can have is 9,344 bytes or 4,672 16-bit words.

In Example 29 we will generate and use a dual-port distributed RAM to form a stack.

Listing 4.4 dist_rom16_top.vhd

```vhdl
-- Example 28: dist_rom16_top
library IEEE;
use IEEE.STD_LOGIC_1164.all;
use work.rom_components.all;

entity dist_rom16_top is
      port(
            mclk : in STD_LOGIC;
            btn : in STD_LOGIC_VECTOR(3 downto 0);
            ld : out STD_LOGIC_VECTOR(7 downto 0);
            a_to_g : out STD_LOGIC_VECTOR(6 downto 0);
            an : out STD_LOGIC_VECTOR(3 downto 0)
         );
end dist_rom16_top;

architecture dist_rom16_top of dist_rom16_top is
signal clk25, clk190, clr, go1, done: std_logic;
signal x: std_logic_vector(15 downto 0);
signal addr: std_logic_vector(3 downto 0);
signal M: std_logic_vector(7 downto 0);
signal gcds, btnd: std_logic_vector(3 downto 0);
begin

clr <= btn(3);
x <= X"000" & gcds;
ld <= M;

U1 : clkdiv
      port map(mclk => mclk, clr => clr, clk25 => clk25,
            clk190 => clk190);

U2 : debounce4
      port map(clk => clk190, clr => clr, inp => btn,
            outp => btnd);

U3 : clock_pulse
      port map(inp => btnd(0), cclk => clk25, clr => clr,
            outp => go1);

U4 : gcd3
      port map(
            clk => clk25, clr => clr, go => go1,
            xin => M(7 downto 4), yin => M(3 downto 0),
            done => done, gcd => gcds);

U5 : x7segb
      port map(x => x, cclk => clk190, clr => clr,
            a_to_g => a_to_g, an => an);

U6 : counter generic map(N => 3)
      port map(clr => clr, clk => go1, q => addr);

U7 : dist_rom16
            port map (a => addr, spo => M);

end dist_rom16_top;
```

Example 29

A Stack

A stack is a group of memory locations in which temporary data can be stored. A stack is different from any other collection of memory locations in that data are put on and taken from the *top* of the stack. The process is similar to stacking dinner plates on top of one another, where the last plate put on the stack is always the first one removed from it. We sometimes refer to this as a *last in-first out* or LIFO stack.

All microcontrollers have a system stack that is used to store the return address from a subroutine call and other state information when interrupts occur. The memory address corresponding to the top of the stack is stored in a register called the *stack pointer*. When data are put on the stack, using a *push* instruction, the stack pointer is *decremented*. This means that the stack grows *backward* in memory. As data values are put on the stack they are put into memory locations with lower addresses. When data are removed from the stack, using a *pop* instruction, the stack pointer is *incremented*. There are two possibilities when pushing data on a stack – the stack pointer can be decremented either *before* or *after* the data value is stored in memory. Both methods have been used in different microprocessors.

In this example we will implement a stack in an FPGA using distributed memory. We will store 16-bit words in the stack and make the stack 32 words deep as shown in Fig. 4.5. If we make *push* = '1' the value $d(15:0)$ will be pushed onto the stack on the rising edge of the clock. The ouput $q(15:0)$ will display the value on the top of the stack. If we make *pop* = '1' the top value will be popped from the stack on the rising edge of the clock and the ouput $q(15:0)$ will display the new value on the top of the stack. If *push* and *pop* are both 0 then no change takes place on the rising edge of the clock. On the other hand if *push* and *pop* are both 1 then on the rising edge of the clock the value of $d(15:0)$ will be stored on the top of the stack without decrementing the stack pointer. We will use this feature in Chapter 9 where this stack will play a central role in our design of a Forth core. If the stack is empty the output *empty* will be 1. If the stack is full the output *full* will be 1.

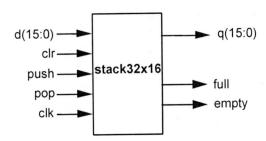

Figure 4.5 A 32 x 16 stack

Inside the *stack32x16* component shown in Fig. 4.5 we will have a 32x16 distributed RAM that we generate from the Core Generator. Our stack pointer register contain the address of the top of the stack. Should this be the address of the last value stored in the stack, or should it be the address of the next available (empty) location? In the former case we would decrement the stack pointer and then store the value at that new address. This has the advantage that the address will always be pointing to the last value stored on the stack and that will be the output value $q(15:0)$ in Fig. 4.5. However, this will have the disadvantage of taking two clock cycles to push a value on the stack – one clock cycle to decrement the stack pointer, and a second clock cycle to write a word to memory. Suppose we choose our second option and write the value at the address stored in the stack pointer and then decrement the stack pointer. This has the advantage of being able to do both operations on the same clock cycle. The disadvantage is that the resulting address will be pointing to the next empty location, so that the output $q(15:0)$ in Fig. 4.5 will not be displaying the value on top of the stack – in fact, it will be displaying garbage! So what should we do?

We want to push a value on the stack in a single clock cycle and at the same time we want the output $q(15:0)$ in Fig. 4.5 to display this value after the stack pointer has been decremented. The way to do this is to use a *dual-port* RAM for the stack in which we can write to one address and read from a second address at the same time. Our stack will therefore look like Fig. 4.6 in which the address *rd_addr* is always equal to *wr_addr* + 1. To push a value on the stack in a single clock cycle we would store the value at the address *wr_addr* and decrement both *wr_addr* and *rd_addr*. At this point *rd_addr* will be pointing to the value just stored in memory and would therefore be the value of the output $q(15:0)$. To pop a value from the stack both *wr_addr* and *rd_addr* are incremented.

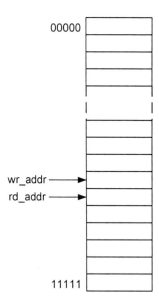

Figure 4.6 Values are written to *wr_addr* and read from *rd_addr*

A circuit diagram for the component *stack32x16* is shown in Fig. 4.7. The dual-port RAM component *dpram32x16* is generated from the Distributed Memory Generator as shown in Fig. 4.8. In this component if *we* = '1' then on the rising edge of the clock, *clk*, the value *d*(15:0) is stored at address *a*(4:0). The output *spo*(15:0) is the value stored at address *a*(4:0) and the output *dpo*(15:0) is the value stored at address *dpra*(4:0).

The multiplexer in Fig. 4.7 allows the write address to be either *wr_addr* or *rd_addr*. It will be *wr_addr* for a push operation (*push* = '1') but will be *rd_addr* if *push* = '1' **and** *pop* = '1'. This is the special case when we don't want to decrement *wr_addr* and *rd_addr*. These operations are summarized in Table 4.1.

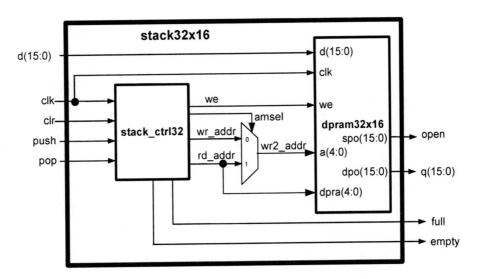

Figure 4.7 Circuit diagram for a 32 x 16 stack

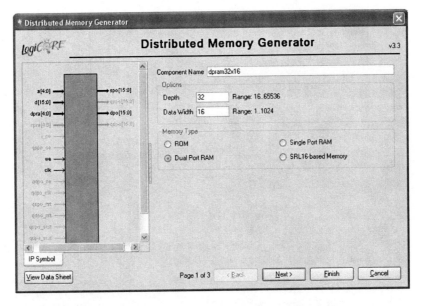

Figure 4.8 Generating the component dpram32x16

Example 29

Table 4.1 Push and Pop Operations

Operation	push	pop	amsel	we
No operation	0	0	0	0
Pop value from stack	0	1	0	0
Push value on stack	1	0	0	1
Write to top of stack	1	1	1	1

The component *stack_ctrl32* shown in Fig. 4.7 will generate the appropriate values for *rd_addr*, *wr_addr*, *amsel*, and *we*. Listing 4.5 is the VHDL program for *stack_ctrl32*. Note that the variables *push_addr* and *pop_address* get assigned to the outputs *wr_addr* and *rd_addr* respectively at the end of the process. The variable *push_addr* is initialized to 11111 and the variable *pop_addr* is initialized to 00000 (which is 1 greater that 11111). When *wr_addr* is 00000 and a value is pushed on the stack, the stack will be full. The variable *push_addr* gets decremented to 11111 which will set *full_flag* to 1 and reset *push_addr* to 00000. When *full_flag* is 1 no further values can be pushed on the stack. Similarly, values can be popped from the stack only when *empty_flag* is 0.

Listing 4.6 is the VHDL program for the *stack32x16* component shown in Fig. 4.7. A simulation of this *stack32x16* program is shown in Fig. 4.9.

Listing 4.5 stack_ctrl32.vhd

```vhdl
-- Example 29a: Stack controller
library IEEE;
use IEEE.STD_LOGIC_1164.all;
use IEEE.STD_LOGIC_unsigned.all;

entity stack_ctrl32 is
    port (
        clr: in STD_LOGIC;
        clk: in STD_LOGIC;
        push: in STD_LOGIC;
        pop: in STD_LOGIC;
        we: out STD_LOGIC;
    amsel: out STD_LOGIC;
        wr_addr: out STD_LOGIC_VECTOR (4 downto 0);
        rd_addr: out STD_LOGIC_VECTOR (4 downto 0);
        full: out STD_LOGIC;
        empty: out STD_LOGIC
    );
end stack_ctrl32;

architecture stack_ctrl32_arch of stack32_ctrl is
signal full_flag, empty_flag: STD_LOGIC;
begin
  stk: process(clr, clk, push, pop, full_flag, empty_flag)
    variable push_addr, pop_addr: STD_LOGIC_VECTOR(4 downto 0);
```

Listing 4.5 (cont.) stack_ctrl32.vhd

```vhdl
  begin
    if clr = '1' then
      push_addr := "11111";
      pop_addr := "00000";
      empty_flag <= '1';
      full_flag <= '0';
      wr_addr <= "11111";
      rd_addr <= "00000";
      full <= full_flag;
      empty <= empty_flag;
    elsif clk'event and clk = '1' then
      if push = '1' then
        if pop = '0' then
          if full_flag = '0' then
            push_addr := push_addr - 1;
            pop_addr := push_addr + 1;
            empty_flag <= '0';
              if push_addr = "11111" then
                full_flag <= '1';
                push_addr := "00000";
              end if;
          end if;
        else
          -- write to top of stack (pop_addr) without pushing
          -- don't change push_addr and pop_addr
        end if;
      elsif pop = '1' then
        if empty_flag = '0' then
          pop_addr := pop_addr + 1;
          if full_flag = '0' then
            push_addr := push_addr + 1;
          end if;
          full_flag <= '0';
          if pop_addr = "00000" then
            empty_flag <= '1';
          end if;
        end if;
      end if;
      wr_addr <= push_addr;
      rd_addr <= pop_addr;
    end if;
    full <= full_flag;
    empty <= empty_flag;
    if push = '1' and full_flag = '0' then
      we <= '1';
    else
      we <= '0';
    end if;
    if push = '1' and pop = '1' then
      amsel <= '1';
    else
      amsel <= '0';
    end if;
  end process stk;
end stack_ctrl32_arch;
```

Listing 4.6 stack32x16.vhd

```vhdl
-- Example 29b: stack32x16
library IEEE;
use IEEE.STD_LOGIC_1164.all;

entity stack32x16 is
    port(
            clk : in STD_LOGIC;
            clr : in STD_LOGIC;
            push : in STD_LOGIC;
            pop : in STD_LOGIC;
            d : in STD_LOGIC_VECTOR(15 downto 0);
            full : out STD_LOGIC;
            empty : out STD_LOGIC;
            q : out STD_LOGIC_VECTOR(15 downto 0)
         );
end stack32x16;

architecture stack32x16 of stack32x16 is

    component dpram32x16
    port (
    a: IN std_logic_VECTOR(4 downto 0);
    d: IN std_logic_VECTOR(15 downto 0);
    dpra: IN std_logic_VECTOR(4 downto 0);
    clk: IN std_logic;
    we: IN std_logic;
    spo: OUT std_logic_VECTOR(15 downto 0);
    dpo: OUT std_logic_VECTOR(15 downto 0));
    end component;

    component stack_ctrl32
    port(
            clr : in std_logic;
            clk : in std_logic;
            push : in std_logic;
            pop : in std_logic;
            we : out std_logic;
            amsel : out std_logic;
            wr_addr : out std_logic_vector(4 downto 0);
            rd_addr : out std_logic_vector(4 downto 0);
            full : out std_logic;
            empty : out std_logic);
    end component;

    component mux2g
    generic(
            N : INTEGER);
    port(
            a : in std_logic_vector(N-1 downto 0);
            b : in std_logic_vector(N-1 downto 0);
            s : in std_logic;
            y : out std_logic_vector(N-1 downto 0));
    end component;
```

Listing 4.6 (cont.) stack32x16.vhd

```vhdl
        signal we, amsel: std_logic;
        signal wr_addr, rd_addr, wr2_addr: std_logic_vector(4 downto 0);
begin

U1 : dpram32x16
            port map (
                    a => wr2_addr,
                    d => d,
                    dpra => rd_addr,
                    clk => clk,
                    we => we,
                    spo => open,
                    dpo => q);

U2 : stack_ctrl32
        port map(
                clr => clr,
                clk => clk,
                push => push,
                pop => pop,
                we => we,
                amsel => amsel,
                wr_addr => wr_addr,
                rd_addr => rd_addr,
                full => full,
                empty => empty
        );

U3 : mux2g generic map(N => 5)
        port map(a => wr_addr, b => rd_addr, s => amsel, y => wr2_addr);

end stack32x16;
```

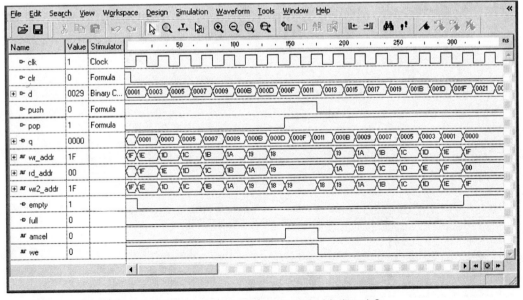

Figure 4.9 Simulation of the stack in Listing 4.6

Example 30

Block RAM

The Xilinx Spartan3E-500 FG320 FPGA used on the Nexys-2 board from Digilent contains 368,640 bits of block RAM/ROM. Thus the maximum amount of block RAM that you can use is 46,080 bytes or 23,040 16-bit words.

To create a block ROM run Core Generator and select *Single Port Block Mem*.

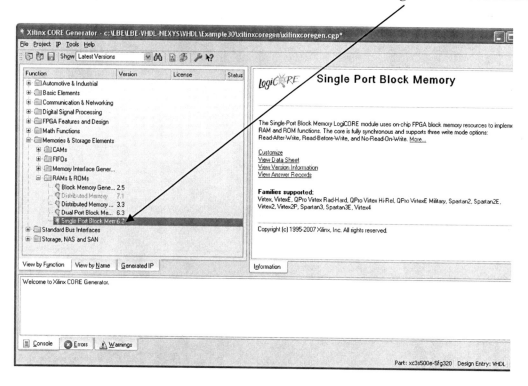

We will create an 8 x 16 block ROM containing the 16-bit data given in the *.coe* file shown in Listing 4.7.

Listing 4.7 Example30.coe
```
; Example 30 Initialization file for a 8x16 block ROM
memory_initialization_radix = 16;
memory_initialization_vector =
0000 1111 2222 3333
4444 5555 6666 7777;
```

Type *brom8x16* for the component name, select *Read Only*, and select a width of 16 and a depth of 8.

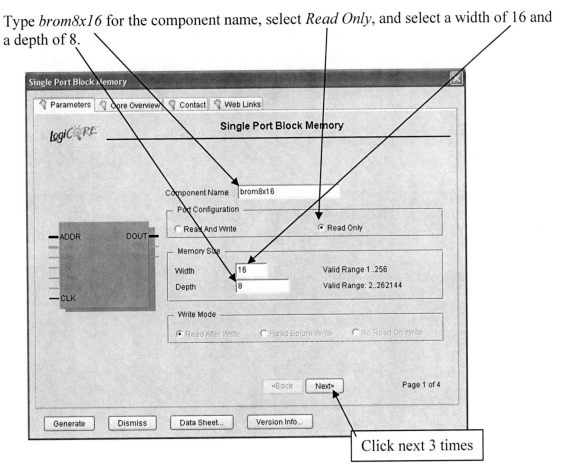

Click next 3 times

Check *Load Init File*

Click *Load File...*

Find *Example30.coe* from Listing 4.7.

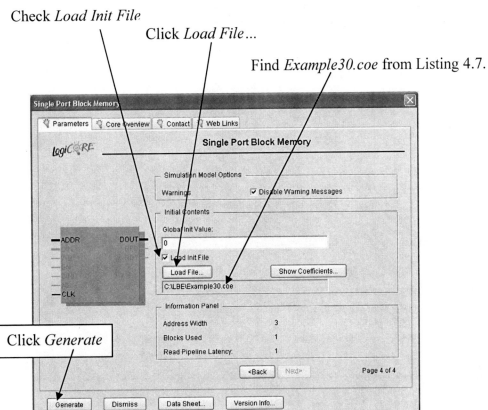

Click *Generate*

Store the component declaration for *brom8x16* from the file *brom8x16.vho* in the *rom_components.vhd* package. Listing 4.8 shows a VHDL program that will display the contents of the ROM on the 7-segment display of the Nexys-2 board as you press *btn*(0). You can copy the U5 instantiation of the component *brom8x16* in Listing 4.8 from the file *brom8x16.vho*. To synthesize this top-level design you must copy the files *brom8x16.edn* and *brom8x16.mif* from the *xilinxcoregen* directory to the *src* directory.

Listing 4.8 brom8x16.vhd

```vhdl
-- Example 30: brom8x16_top
library IEEE;
use IEEE.STD_LOGIC_1164.all;
use work.rom_components.all;

entity brom8x16_top is
      port(
            mclk  : in STD_LOGIC;
            btn   : in STD_LOGIC_VECTOR(3 downto 0);
            ld    : out STD_LOGIC_VECTOR(7 downto 0);
            a_to_g : out STD_LOGIC_VECTOR(6 downto 0);
            an    : out STD_LOGIC_VECTOR(3 downto 0)
          );
end brom8x16_top;

architecture brom8x16_top of brom8x16_top is

signal clkp, clk190, clr: std_logic;
signal x: std_logic_vector(15 downto 0);
signal addr: std_logic_vector(2 downto 0);
begin

clr <= btn(3);
ld <= "00000" & addr;

U1 : clkdiv
      port map(mclk => mclk, clr => clr, clk190 => clk190);

U2 : clock_pulse
      port map(inp => btn(0), cclk => clk190, clr => clr,
               outp => clkp);

U3 : x7segb
      port map(x => x, cclk => clk190, clr => clr,
               a_to_g => a_to_g, an => an);

U4 : counter generic map(N => 3)
      port map(clr => clr, clk => clkp, q => addr);

U5 : brom8x16
         port map (addr => addr, clk => clkp, dout => x);

end brom8x16_top;
```

Example 31

External RAM

The Nexys-2 board contains two onboard memory devices that are external to the FPGA. The first is a 16 Mbyte RAM organized as 8 Meg x 16-bit words as shown in Fig. 4.10. The second is the same size flash memory as shown in Fig. 4.11. In this example we will show how to interface the FPGA to the external RAM shown in Fig. 4.10. We will show how to use the flash memory in Example 32.

The RAM and flash memories share a common 23-bit address bus $A(22:0)$ and 16-bit bi-directional data bus $DQ(15:0)$ as well as common write enable ($WE\#$) and output enable ($OE\#$) input signals. The # in these signal names indicate that these are active low inputs. We will call the corresponding outputs from the FPGA WE_L and OE_L to indicate that they are active low signals. The other output and input names associated with the RAM that are used in the *nexys2.ucf* file are shown in Fig. 4.10.

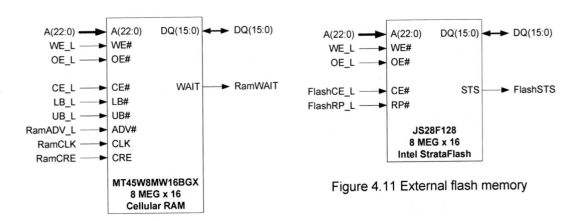

Figure 4.10 External cellular RAM

Figure 4.11 External flash memory

The cellular RAM shown in Fig. 4.10 can operate in a number of modes that you can read about by downloading its datasheet from the web (by searching for its part number). The default mode is an industry standard asynchronous read and write operation. In this mode the inputs *RamADV_L*, *RamCLK*, and *RamCRE* can all be set to zero and the output *RamWAIT* can be ignored. The RAM is selected by bringing the chip enable input CE_L low. You should also make sure that the flash memory is disabled by setting *FlashCE_L* high.

You can read or write either the upper byte $DQ(15:8)$ if UB_L is low, the lower byte $DQ(7:0)$ if LB_L is low, or both bytes $DQ(15:0)$ if both UB_L and LB_L are low. We will always read or write 16-bits at a time and therefore will set UB_L and LB_L to zero.

To read a 16-bit value from memory you would put the address you want to read from on the address bus $A(22:0)$, set the write enable WE_L high, and set the output enable OE_L low. After waiting the memory access time – 70 ns – the valid data will show up on the data bus $DQ(15:0)$. The 70 ns access time will limit the clock frequency that we can use to read data from the memory. We won't be able to change the read address faster than every 80 ns which is a clock frequency of 12.5 MHz, which can be obtained from $q(1)$ in our clock divide component *clkdiv*.

To write a 16-bit value to memory you would put the address you want to write to on the address bus $A(22:0)$ and the data you want to write on the data bus $DQ(15:0)$. You would then bring the write enable WE_L low for at least 45 ns. When you bring WE_L back high the data will be latched into the memory at the address $A(22:0)$. This address must remain valid for at least 70 ns, and therefore the maximum speed that we can write data to memory will be 12.5 MHz. Normally, you will set OE_L high when writing data to memory, although its value is actually "Don't Care" and the WE_L signal will override it.

Because the data bus $DQ(15:0)$ is a bi-directional bus we must use a tri-state buffer to interface it to the FPGA. The component *buff3* in the top-level design in Fig. 4.12 shows how we do this.

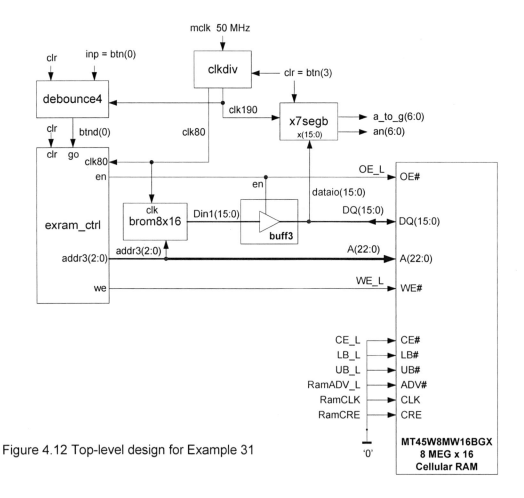

Figure 4.12 Top-level design for Example 31

In the component *buff3* if *en* = '1' then the output of *buff3* is equal to its input $Din1(15{:}0)$. This would be the situation when you want to write the value $Din1(15{:}0)$ to the memory. On the other hand, if *en* = '0' then the output of *buff3* is high impedance – the equivalent of an open circuit. In this case the ouput *OE_L* will also be low enabling the output of the memory $DQ(15{:}0)$, which will show up as an input to the *x7segb* component. This would be the situation when you want to read a 16-bit value from memory and display it on the 7-segment display. Listing 4.9 shows the VHDL program for the component *buff3*. Note the use of the *when...else* statement to implement the tri-state buffer.

Listing 4.9 buff3.vhd
```
-- Example 31a: Tri-state buffer
library IEEE;
use IEEE.STD_LOGIC_1164.all;

entity buff3 is
    generic (N:integer);
    port(
            input  : in STD_LOGIC_vector(N-1 downto 0);
            en     : in STD_LOGIC;
            output : out STD_LOGIC_vector(N-1 downto 0)
        );
end buff3;


architecture buff3 of buff3 is
begin
    output <= input when en = '1' else (others => 'Z');
end buff3;
```

The top-level design in Fig. 4.12 will write the eight 16-bit values stored in *brom8x16* (see Listing 4.7 of Example 30) to the external RAM when you press *btn*(0). If you continue to press *btn*(0) each of the values stored in the RAM will be read, one at at time, and displayed on the 7-segment display.

The component *exram_ctrl* in Fig. 4.12 implements the state diagram shown in Fig. 4.13. This state machine uses the 12.5 MHz clock (*clk*80) with an 80 ns period. When the *go* signal goes high the first time (by pressing *btn*(0)) the state diagram goes around the loop on the right eight times – getting the next address, reading the 16-bit value from *brom8x16*, writing the value to the RAM, and checking to see if the address has incremented back to 000. It then goes to state *wtngo*, which will wait for the *go* signal to go low, i.e. for you to lift your finger from *btn*(0). The *read* state will wait for you to press *btn*(0) again while displaying on the 7-segment display the current value being read (from address 000). When you press *btn*(0) again the address will be incremented and the state diagram in Fig. 4.13 goes around the loop on the left eight times – checking to see if the address has incremented back to 000, waiting for you to lift your finger, and then waiting for you to press *btn*(0) again while displaying the current value on the 7-segment display. After displaying all eight values the state diagram returns to the *start* state and you can just keep going. Listing 4.10 shows the VHDL program for the component *exram_ctrl* and its simulation is shown in Fig. 4.14.

122   Example 31

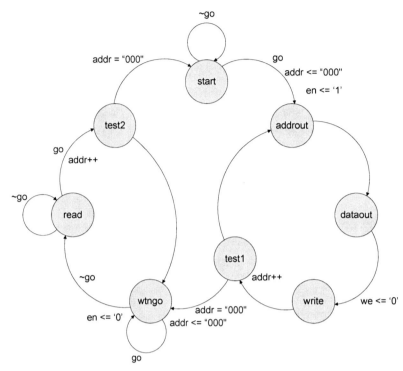

Figure 4.13 State diagram for the component *exram_ctrl*

Listing 4.10 exram_ctrl.vhd

```vhdl
-- Example 31b: exram_ctrl
library IEEE;
use IEEE.STD_LOGIC_1164.all;
use IEEE.STD_LOGIC_unsigned.all;

entity exram_ctrl is
        port(
                clk80 : in STD_LOGIC;
                clr : in STD_LOGIC;
                go : in STD_LOGIC;
                we : out STD_LOGIC;
                en : out STD_LOGIC;
                addr3 : out STD_LOGIC_VECTOR(2 downto 0)
             );
end exram_ctrl;

architecture exram_ctrl of exram_ctrl is
type state_type is (start, addrout, dataout, write, test1,
                    wtngo, read, test2);
signal state: state_type;
begin
sreg: process(clk80, clr)
variable addrv: STD_LOGIC_VECTOR(2 downto 0);
begin
  if clr = '1' then
      state <= start;
      addrv := (others => '0');
      we <= '1';
      en <= '0';
   elsif clk80'event and clk80 = '1' then
```

Listing 4.10 (cont.) exram_ctrl.vhd

```vhdl
    case state is
        when start =>
            we <= '1';
            if go = '1' then
                state <= addrout;
                addrv := "000";
                en <= '1';
            else
                state <= start;
            end if;
        when addrout =>
            state <= dataout;
            we <= '1';
        when dataout =>
           state <= write;
           we <= '0';
        when write =>
           state <= test1;
           we <= '1';
        when test1 =>
            we <= '1';
            addrv := addrv + 1;
            if addrv = "000" then
                  state <= wtngo;
                  en <= '0';
            else
                  state <= addrout;
            end if;
        when wtngo =>
            we <= '1';
            if go = '1' then
                state <= wtngo;
            else
                state <= read;
            end if;
        when read =>
            we <= '1';
            if go = '1' then
                state <= test2;
                    addrv := addrv + 1;
            else
                state <= read;
            end if;
        when test2 =>
            we <= '1';
            if addrv = "000" then
                  state <= start;
            else
                  state <= wtngo;
            end if;
        when others =>
            null;
    end case;
  end if;
  addr3 <= addrv;
end process;
end exram_ctrl;
```

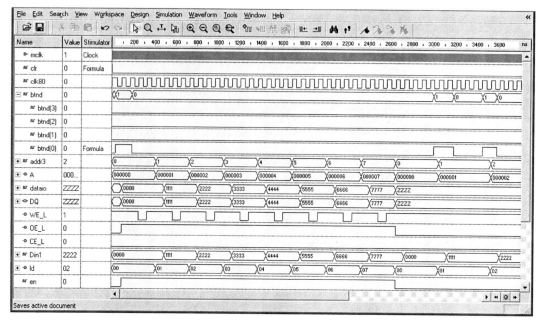

Figure 4.14 Simulation of the component *exram_ctrl* in Listing 4.10

Listing 4.11 shows the VHDL program for the top-level design in Fig. 4.12. Note in the entity that the bi-directional *DQ* is of type *inout*. In the architecture it is assigned to the signal *dataio*, which is connected to the output of the tri-state buffer *buff3*.

Listing 4.11 ext_ram_top.vhd

```vhdl
-- Example 31c: ext_ram_top
library IEEE;
use IEEE.STD_LOGIC_1164.all;
use work.rom_components.all;

entity ext_ram_top is
    port(
        mclk : in STD_LOGIC;
        btn : in STD_LOGIC_VECTOR(3 downto 0);
        ld : out STD_LOGIC_VECTOR(7 downto 0);
        a_to_g : out STD_LOGIC_VECTOR(6 downto 0);
        an : out STD_LOGIC_VECTOR(3 downto 0);
        A : out STD_LOGIC_VECTOR(22 downto 0);
        DQ : inout STD_LOGIC_VECTOR(15 downto 0);
        CE_L : out STD_LOGIC;
        UB_L : out STD_LOGIC;
        LB_L : out STD_LOGIC;
        WE_L : out STD_LOGIC;
        OE_L : out STD_LOGIC;
        FlashCE_L : out STD_LOGIC;
        RamCLK : out STD_LOGIC;
        RamADV_L : out STD_LOGIC;
        RamCRE : out STD_LOGIC
        );
end ext_ram_top;
```

Listing 4.11 (cont.) ext_ram_top.vhd

```vhdl
architecture ext_ram_top of ext_ram_top is

signal clk80, clk190, clr, en: std_logic;
signal btnd: std_logic_vector(3 downto 0);
signal addr3: std_logic_vector(2 downto 0);
signal Din1, dataio: std_logic_vector(15 downto 0);
begin

      clr <= btn(3);
      ld <= "00000" & addr3;
      A <= "000000000000000000" & addr3;
      FlashCE_L <= '1';        -- Disable Flash
      CE_L <= '0';             -- Enable ram
      UB_L <= '0';
      LB_L <= '0';
      RamCLK <= '0';
      RamADV_L <= '0';
      RamCRE <= '0';
      OE_L <= en;
      DQ <= dataio;

U1 : clkdiv
      port map(mclk => mclk, clr => clr, clk190 => clk190,
            clk80 => clk80);

U2 : debounce4
      port map(cclk => clk190, clr => clr, inp => btn,
            outp => btnd);

U3 : x7segb
      port map(x => dataio, cclk => clk190, clr => clr,
            a_to_g => a_to_g, an => an);

U4 : exram_ctrl
      port map(clk80 => clk80, clr => clr, go => btnd(0),
            we => WE_L, en => en, addr3 => addr3);

U5 : brom8x16
      port map (addr => addr3, clk => clk80, dout => Din1);

U6 : buff3 generic map(N => 16)
      port map(input => Din1, en => en, output => dataio);

end ext_ram_top;
```

Example 32

External Flash Memory

In Example 31 we saw that the external RAM and flash memories share a common 23-bit address bus $A(22:0)$ and 16-bit bi-directional data bus $DQ(15:0)$ as well as common write enable ($WE\#$) and output enable ($OE\#$) input signals. We also saw in Example 31 how to write data to the RAM and read it back. The RAM is a volatile memory, which means when you turn off the power the data in the RAM goes away. A flash memory, on the other hand, is a non-volatile memory, which means that the data remains in the memory when you turn the power off. This is useful for many types of data that you would like to have available when you power-up your FPGA board.

Writing data to a flash memory is more complicated than writing data to a RAM. You actually have to program the flash memory. The easiest way to do this on the Nexys-2 board is to use the *MemUtil* application program available from Digilent at http://www.digilentinc.com/Software/MMU.cfm?Nav1=Software&Nav2=MMU. This program requires that you have already installed the *Adept Suite* software that includes the *ExPort* program we have been using.

We will illustrate the use of the *MemUtil* software by loading (programming) the flash memory with the four 16-bit hex numbers shown in Listing 4.12. We have simply written these four hex numbers in a file called *hex4.txt* using *Notepad*. This text file will actually contain the ASCII codes of the characters that we typed. Thus, if you looked to see what was really stored in the file you would find the following hex bytes:

```
31 32 33 34 0D 0A 35 36 37 38 0D 0A
39 30 41 42 0D 0A 43 44 45 46 0D 0A
```

The hex value 0D is the ASCII code for a *carriage return* (CR) and 0A is the ASCII code for a *line feed* (LF). If we loaded the file *hex4.txt* into the flash memory, it would be these ASCII codes that get stored in the flash memory.

Listing 4.12 hex4.txt

```
1234
5678
90AB
CDEF
```

If we really want to store the data shown in Listing 4.12 as four hex digits (a total of eight bytes) rather than the 24 bytes of ASCII codes shown above, we need to convert the ASCII text file to a corresponding binary file containing only eight bytes. The Matlab function *asc2bin(infile, outfile)* shown in Listing 4.13 will do this. Fig. 4.15 shows a sample run that creates the binary file *hex4.out*.

Listing 4.13 asc2bin.m

```
function asc2bin(infile, outfile)
% Input a .txt number file containing ascii hex numbers
% and save it as a binary file (.out)
% for loading to flash memory
% asc2bin(infile, outfile)
% Example:
% asc2bin('hex4.txt', 'hex4.out');

fid1 = fopen(infile,'r');         %opens the input file to read
fid2 = fopen(outfile,'w');        %opens the output file to write
A = fscanf(fid1,'%x',inf);        %read input file
fwrite(fid2,uint16(A),'uint16')   %write output file
fclose(fid1);                     %close input file
fclose(fid2);                     %close output file
```

```
>> help asc2bin
  Input a .txt number file containing ascii hex numbers
  and save it as a binary file (.out)
  for loading to flash memory
  asc2bin(infile, outfile)
  Example:
  asc2bin('hex4.txt', 'hex4.out');

>> asc2bin('hex4.txt', 'hex4.out')
```

Figure 4.15 Example of Matlab run to create binary file *hex4.out*

To load the eight bytes from the file *hex4.out* into the flash memory you must first program the FPGA with the bit file *onboardmemcfgJtagClk.bit* that you get when you download *MemUtil* from Digilent. You would then run *MemUtil* and click on the *Load Flash* tab as shown in Fig. 4.16. Enter the *File Name*, type 0 for the *File Start Location*, type 8 for the *Length*, and click *Load*. That's it – your flash memory now contains the four 16-bit hex numbers shown in Listing 4.12.

Reading the memory is now easy – at least if we do it slowly enough. The top-level design shown in Fig. 4.17 will allow us to read out the four words – one at a time by pressing *btn*(0) – and display the results on the 7-segment display.

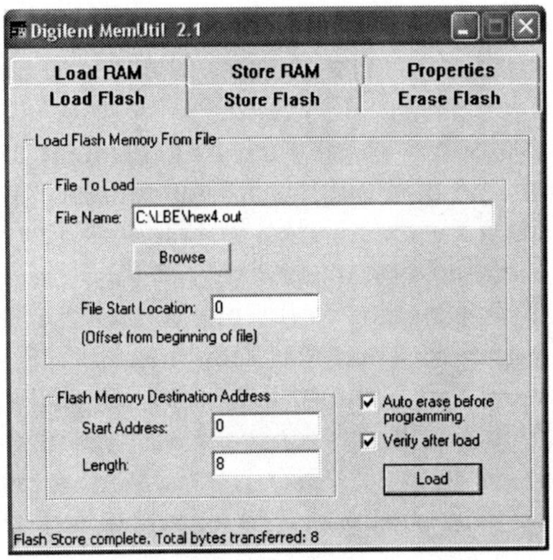

Figure 4.16
Using *MemUtil* to load eight bytes from *hex4.out* into the flash memory

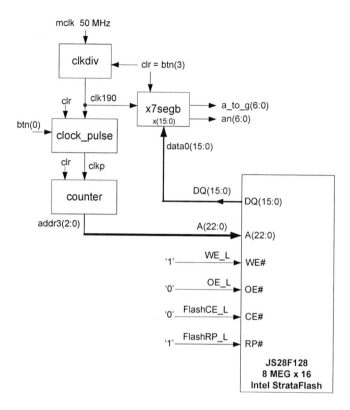

Figure 4.17 Top-level design for reading data from the flash memory

Listing 4.14 is a VHDL program that implements the top-level design shown in Fig. 4.17. Note that because we are only reading data from the flash memory we can make *DQ* in the entity only an input. We use a 3-bit counter to generate the eight addresses 000 – 111 and concatenate this address with leading zeros to form the flash address *A*(22:0). The counter clock is a clock pulse generated every time you press *btn*(0).

Note that 16-bit data are contained in only the first four addresses (0 – 3) even though the counter outputs a 3-bit address. When you run the program addresses (0 – 3) will contain the data shown in Listing 4.12 while addresses (4 – 7) will contain 0xFFFF from the erased flash memory.

This program works fine in reading the flash data one word at a time. How fast can your program read the flash memory? To read the first word will take up to 120 ns. However, if this address is 0, it will actually buffer the four words at addresses 0, 1, 2, and 3. You can then access each of these next three words by waiting only 25 ns between reads rather than 120 ns. If you then access a word at another address on a 4-word boundary, you must wait 120 ns, which will then buffer the next four words. In Example 40 we will see how we must use this feature in order to read the contents of the flash memory fast enough to display the data on a VGA video monitor.

Listing 4.14 flash_top.vhd

```vhdl
-- Example 32: flash_top
library IEEE;
use IEEE.STD_LOGIC_1164.all;
use work.rom_components.all;

entity flash_top is
  port(
        mclk : in STD_LOGIC;
        btn : in STD_LOGIC_VECTOR(3 downto 0);
        ld : out STD_LOGIC_VECTOR(7 downto 0);
        a_to_g : out STD_LOGIC_VECTOR(6 downto 0);
        an : out STD_LOGIC_VECTOR(3 downto 0);
        A : out STD_LOGIC_VECTOR(22 downto 0);
        DQ : in STD_LOGIC_VECTOR(15 downto 0);
        CE_L : out STD_LOGIC;
        WE_L : out STD_LOGIC;
        OE_L : out STD_LOGIC;
        FlashCE_L : out STD_LOGIC;
        FlashRP_L : out STD_LOGIC
       );
end flash_top;

architecture flash_top of flash_top is

signal clkp, clk190, clr: std_logic;
signal data0: std_logic_vector(15 downto 0);
signal addr3: std_logic_vector(2 downto 0);
begin

  clr <= btn(3);
  ld <= "00000" & addr3;
  A <= "00000000000000000000" & addr3;
  FlashCE_L <= '0';         -- Enable Flash
  CE_L <= '1';              -- Disable ram
  WE_L <= '1';
  FlashRP_L <= '1';
  OE_L <= '0';
  data0 <= DQ;


U1 : clkdiv
  port map(mclk => mclk, clr => clr, clk190 => clk190);

U2 : clock_pulse
  port map(inp => btn(0), cclk => clk190, clr => clr,
           outp => clkp);

U3 : x7segb
  port map(x => data0, cclk => clk190, clr => clr,
           a_to_g => a_to_g, an => an);

U4 : countergeneric map(N => 3)
  port map(clr => clr, clk => clkp, q => addr3);

end flash_top;
```

5

UART

In this chapter we will show how to implement a component for transmitting and receiving data through a standard RS-232 serial port in VHDL.

Serial communication requires only two signals, *TxD* and *RxD*. The former refers to transmitted data and the latter to received data. A standard 9-pin serial port found on a personal computer is usually a male connector with pin 2 for *RxD* and pin 3 for *TxD*. On the other hand, a 9-pin serial port found on the Nexys-2 board is a female connector that uses pin 2 as *TxD* and pin 3 as *RxD*. When connected, the *TxD* of one is connected to the *RxD* of the other and vice versa so that data transmitted by one is received by the other. Data is transferred at a specific rate known as a *baud rate* given in bits per second. Common baud rates include 4800, 9600, 56k, and 115.2k. In addition to the baud rate, the number of data bits, parity, and number of stop bits must be selected and match in both the transmitting and receiving devices. Most implementations transfer from five to eight data bits with even, odd, or no parity including one or two stop bits. For simplicity, we will transfer information using eight data bits, no parity, and one stop bit.

Serial transmission is synchronized using a start bit preceding the transmission and a stop bit indicating that the last bit has been transmitted completing the byte transfer. ASCII codes are often used for communicating between a PC and a device or between two devices. The standard ASCII codes are shown in Table 5.1. Figure 5.1 shows the bit sequence for transmitting the ASCII code for a capital "T", 0x54.

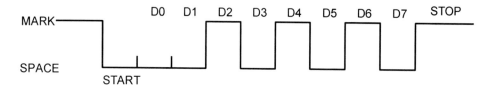

Figure 5.1 ASCII code 0x54 = 0101 0100 ("T") sent with no parity

Common baud rates and associated timing information is shown in Fig. 5.2. We will use a 9600 baud rate in Example 33. At 9600 baud (bits per second), each bit is present on the transmit or receive line for 0.104 milliseconds. A character can be transmitted in 1.04 milliseconds since ten bits are required to transmit a single byte, a start bit, eight bits of data, plus a stop bit. This transmission rate yields 960 characters per second.

Table 5.1 Standard ASCII Codes

Dec →		0	16	32	48	64	80	96	112
↓	Hex	0	1	2	3	4	5	6	7
0	0	NUL	DLE	Blank	0	@	P		p
1	1	SOH	DC1	!	1	A	Q	a	q
2	2	STX	DC2	"	2	B	R	b	r
3	3	ETX	DC3	#	3	C	S	c	s
4	4	EOT	DC4	$	4	D	T	d	t
5	5	ENQ	NAK	%	5	E	U	e	u
6	6	ACK	SYN	&	6	F	V	f	v
7	7	BEL	ETB	'	7	G	W	g	w
8	8	BS	CAN	(	8	H	X	h	x
9	9	HT	EM	)	9	I	Y	i	y
10	A	LF	SUB	*	:	J	Z	j	z
11	B	VT	ESC	+	;	K	[	k	{
12	C	FF	FS	,	<	L	\	l		
13	D	CR	GS	-	=	M	]	m	}
14	E	SO	RS	.	>	N	^	n	~
15	F	SI	US	/	?	O	_	o	DEL

Baud rate	Bit time (msec)	No. of STOP bits	Char. time (msec.)	Char./sec.
110	9.09	2	100.00	10
300	3.33	1	33.33	30
600	1.67	1	16.67	60
1200	0.833	1	8.33	120
2400	0.417	1	4.17	240
4800	0.208	1	2.08	480
9600	0.104	1	1.04	960
14400	0.069	1	0.69	1440
19200	0.052	1	0.52	1920
28800	0.035	1	0.35	2880
38400	0.026	1	0.26	3840

Figure 5.2 Common Asynchronous Serial Baud Rates

Example 33

Transmit Module

In this example, we will design a component, *uart_tx*, for transferring data using serial communication. Fig. 5.3 shows the entity for this component. The component *uart_tx* has *clr* and *clk* inputs used to reset and synchronize communication. A byte of data is input using *tx_data*(7:0). When *ready* is asserted high, the byte of data is transmitted on the *TxD* output starting with the least significant bit first. After the transmission has completed, the transmit data ready pin, *tdre*, goes high.

Transmission begins with the *TxD* line transitioning from high to low for one bit time. This leading bit is called the *start bit* (see Fig. 5.1). The bit time depends on the baud rate. Immediately following the

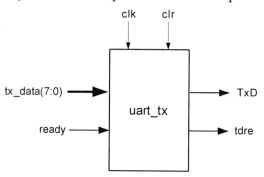

Figure 5.3 *Uart_tx*

start bit, the first data bit, the least significant bit, is transferred followed by the next, more significant bit until all eight bits of data have been transferred. Each bit remains on the *TxD* line for one bit time. After the most significant bit has been transferred, *TxD* goes high for one bit time. This trailing bit is called the *stop bit*.

The state diagram for transmitting serial data is shown in Fig. 5.4.

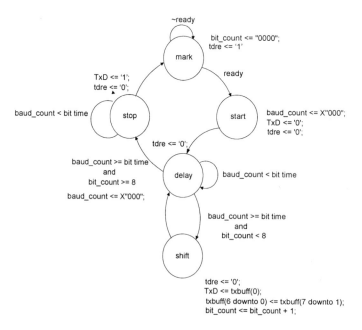

Figure 5.4 State Diagram for Transmitting Serial Data

The state machine starts in the *mark* state until the *ready* signal goes high. This state resets a signal for counting the number of bits transmitted, *bit_count*, to zero and asserts the *tdre* output high indicating that the component is not currently transmitting data. When the *ready* input goes high the state machine transitions to the *start* state for transmitting the start bit. Since the start bit is a logic low, *TxD* is set to zero, and must be held low for one bit time. A counter *baud_count* is used to count clock cycles until the bit time is reached. *Baud_count* is reset to zero in *start* and *tdre* is brought low indicating that a transmission is in progress.

On the rising edge of the clock, the next state is *delay*. The bit time counter, *baud_count* is incremented in the *delay* state and the state machine remains in the *delay* state while the *baud_count* is less than the *bit_time*. *Bit_time* is a constant number of clock cycles required for 0.104 milliseconds to pass. Once the *baud_count* has counted up to the *bit_time*, execution continues to the *shift* state to transfer a data bit. The byte to be transferred is stored in a buffer signal, *txbuff(7:0)*. In this case, the first data bit or least significant bit of the buffer, *txbuff(0)*, is assigned to *TxD* in the *shift* state. Additionally, *tdre* remains low, *txbuff* is shifted one bit to the right, *bit_count* is incremented to count the number of bits transmitted, and *baud_count* is reset to zero. By shifting *txbuff* one bit to the right, *txbuff(0)* always contains the next bit to be transferred until all eight bits have been transferred. On the next rising edge of the clock, the state machine transitions back to *delay*. Once again, *baud_count* is increment on each rising clock edge remaining in *delay* until one bit time has passed, when *baud_count* becomes equal to *bit_time*. Execution continues to *shift*, outputting the next bit and shifting the transfer buffer, *txbuff*. After *TxD* has been set to the next data bit in the *shift* state, the output remains the same while the state machine remains in the *delay* state for one bit time.

This transition between *shift* and *delay* continues incrementing *bit_count* during every *shift* until eight bits have been transmitted. After the eight data bits have been transferred, the state machine transitions to *stop* to transfer the stop bit. To transfer the stop bit, *TxD* is set high. Execution continues in the *stop* state incrementing *baud_count* for one bit time. Once *baud_count* reaches bit time, the state machine transitions to *mark* again setting *tdre* high indicating that the transfer is complete.

Using a 25 MHz clock to drive *uart_tx*, the *bit_time* is computed as follows.

$$25 \times 10^6 \times 0.104 \times 10^{-3} = 2600 \quad (0xA28 \text{ in hex})$$

Listing 5.1 shows the VHDL implementation of *uart_tx*. The *state_type* enumerates the five states. Instead of implementing the state machine as a Moore Machine with three separate processes, we can implement it as a canonical sequential network with a single process. Therefore, only a single *state* variable is necessary. The *baud_count* signal is a 12-bit vector since it will count up to *bit_time*, 0xA28. The 4-bit signal *bit_count* will count to eight.

If the asynchronous clear, *clr*, is high the *state* is set to *mark* and the transfer buffer, baud counter, and bit counter are cleared. *TxD* is set high. On the rising edge of the clock, the *case* statement sets the *state* and operations to be performed.

During the *mark* state, *bit_count* is reset to zero and *tdre* is set high. If *ready* is low, then *mark* will remain the *state* and the 8-bit data input *tx_data* will be latched into the transfer buffer, *txbuff*. If *ready* is high, then *baud_count* is reset to zero and the *state* is set to *start*. In *start*, *TxD* is asserted low and execution continues to *delay*.

The state remains *delay* until *baud_count* is greater than or equal to *bit_time*. Accordingly, *baud_count* is incremented in *delay*. In *shift*, *TxD* is set to the least significant bit of the transfer buffer, *txbuff*(0). The bit counter, *bit_count* is incremented and *txbuff* is shifted one bit to the right. The *stop* state asserts *TxD* high and the *state* remains *stop* until *baud_count* is greater than or equal to *bit_time*. Accordingly, *baud_count* is incremented in *stop*.

Listing 5.1 uart_tx.vhd

```vhdl
-- Example 33a: UART transmitter
library IEEE;
use IEEE.STD_LOGIC_1164.ALL;
use IEEE.STD_LOGIC_UNSIGNED.ALL;

entity uart_tx is
    port(
            clk : in STD_LOGIC;
            clr : in STD_LOGIC;
            tx_data : in STD_LOGIC_VECTOR(7 downto 0);
            ready : in STD_LOGIC;
            tdre : out STD_LOGIC;
            TxD : out STD_LOGIC
        );
end uart_tx;

architecture uart_tx of uart_tx is

type state_type is (mark, start, delay, shift, stop);

signal state: state_type;
signal txbuff: STD_LOGIC_VECTOR (7 downto 0);
signal baud_count: STD_LOGIC_VECTOR (11 downto 0);
signal bit_count: STD_LOGIC_VECTOR (3 downto 0);

constant bit_time: STD_LOGIC_VECTOR (11 downto 0) := X"A28";

begin

uart2: process(clk, clr, ready)
  begin
    if clr = '1' then
      state <= mark;
      txbuff <= "00000000";
      baud_count <= X"000";
      bit_count <= "0000";
      TxD <= '1';
```

Listing 5.1 (cont.) uart_tx.vhd

```vhdl
      elsif (clk'event and clk = '1') then
      case state is
          when mark =>                     -- wait for ready
              bit_count <= "0000";
              tdre <= '1';
              if ready = '0' then
                  state <= mark;
                  txbuff <= tx_data;
              else
                  baud_count <= X"000";
                  state <= start;          --    go to start
              end if;
          when start =>        -- output start bit
              baud_count <= X"000";
              TxD <= '0';
              tdre <= '0';
              state <= delay;    --    go to delay
          when delay =>          -- wait bit time
              tdre <= '0';
              if baud_count >= bit_time then
                  baud_count <= X"000";
                  if bit_count < 8 then    -- if not done
                      state <= shift;   --   go to shift
                  else                  -- else
                      state <= stop;    --   go to stop
                  end if;
              else
                  baud_count <= baud_count + 1;
                  state <= delay;           --    stay in delay
              end if;
          when shift =>            -- get next bit
              tdre <= '0';
              TxD <= txbuff(0);
              txbuff(6 downto 0) <= txbuff(7 downto 1);
              bit_count <= bit_count + 1;
              state <= delay;

          when stop =>              -- stop bit
             tdre <='0';
             TxD <= '1';
              if baud_count >= bit_time then
                 baud_count <= X"000";
                 state <= mark;
              else
                 baud_count <= baud_count + 1;
                 state <= stop;
              end if;
       end case;
     end if;
  end process uart2;
end uart_tx;
```

Testing *Uart_tx*

We can test the *uart_tx* component by adding a separate controller to set *ready* high upon a button press after waiting until a transmission in progress has completed. Fig. 5.5 shows a top-level design for doing this. The *uart_tx* component uses the eight switches to obtain the byte to transmit. The state diagram for *test_tx_ctrl* is shown in Fig. 5.6 and starts by waiting for *btn*(0) to be pressed in the *wtgo* state where the *ready* signal is asserted low for *uart_tx*. Once *btn*(0) is pressed, the state diagram transitions to *wttdre* and remains in *wttdre* until the *tdre* signal goes high. Once the *uart_tx* has finished completing any transfers that might be in progress, *test_tx_ctrl* transitions to the *load* state where *ready* is asserted high to begin transmitting the data set on the switches. Finally, *test_tx_ctrl* waits for the button to return to rest in *wtngo*. This process is repeated to transmit the next byte.

Listing 5.2 shows the VHDL code for the *test_tx_ctrl* controller. Listing 5.3 is a VHDL program that implements the top-level design shown in Fig. 5.5.

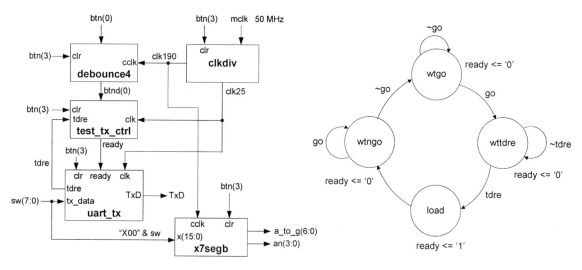

Figure 5.5 Top-level design for Testing *uart_tx*   Figure 5.6 State diagram for *test_tx_ctrl*

Listing 5.2 test_tx_ctrl.vhd

```vhdl
-- Example 33b: Controller for Testing UART transmitter
library IEEE;
use IEEE.STD_LOGIC_1164.all;

entity test_tx_ctrl is
    port(
            clk   : in  STD_LOGIC;
            clr   : in  STD_LOGIC;
            go    : in  STD_LOGIC;
            tdre  : in  STD_LOGIC;
            ready : out STD_LOGIC
        );
end test_tx_ctrl;
```

Listing 5.2 (cont.) test_tx_ctrl.vhd

```vhdl
architecture test_tx_ctrl of test_tx_ctrl is
type state_type is (wtgo, wttdre, load, wtngo);
signal state: state_type;
begin
ctrl: process(clk, clr)
  begin
    if clr = '1' then
      state <= wtgo;
      ready <= '0';
    elsif (clk'event and clk = '1') then
      case state is
          when wtgo =>              -- wait for btn
              if go = '0' then      -- if btn up
                  state <= wtgo;    --    stay in wtgo
                  ready <= '0';
              else
                  ready <= '0';
                  state <= wttdre;  -- else go to wttdre
              end if;
          when wttdre =>            -- wait for tdre = 1
              if tdre = '0' then    -- if tdre = 0
                  state <= wttdre;  --    stay in wtdone
                  ready <= '0';
              else
                  state <= load;    -- else go to load
                  ready <= '0';
              end if;
          when load =>              -- output ready
              ready <= '1';
              state <= wtngo;       -- go to wtngo
          when wtngo =>             -- wait for btn up
              if go = '1' then      -- if btn down
                  state <= wtngo;   --    stay in wtngo
                  ready <= '0';
              else
                  ready <= '0';
                  state <= wtgo;    -- else go to wtgo
              end if;
      end case;
    end if;
  end process ctrl;
end test_tx_ctrl;
```

After implementing the program shown in Listing 5.3 and downloading the bit file to the Nexys-2 board you can set any ASCII code using the switches and then press *btn*(0). This will send the ASCII code out the serial port. To see the result you can run any terminal program running at 9600 baud, such as the program *host.exe* available for download from www.lbebooks.com. For example, if you run *host.exe*, set the switches to hex 41, and press *btn*(0) then the upper-case letter *A* will appear on the *host.exe* screen.

Listing 5.3 uart_tx_top.vhd

```vhdl
-- Example 33c: Testing UART transmitter
library IEEE;
use IEEE.STD_LOGIC_1164.all;
use work.uart_components.all;

entity uart_tx_top is
    port(
            mclk : in STD_LOGIC;
            btn : in STD_LOGIC_VECTOR(3 downto 0);
            sw : in STD_LOGIC_VECTOR(7 downto 0);
            TxD : out STD_LOGIC;
            dp : out STD_LOGIC;
            a_to_g : out STD_LOGIC_VECTOR(6 downto 0);
            an : out STD_LOGIC_VECTOR(3 downto 0)
        );
end uart_tx_top;

architecture uart_tx_top of uart_tx_top is

signal clk25, clk190, clr, tdre, ready: std_logic;
signal x: std_logic_vector(15 downto 0);
signal btnd: std_logic_vector(3 downto 0);
begin

clr <= btn(3);
x <= X"00" & sw;
dp <= '1';       -- decimal point off

U1 : clkdiv
    port map(mclk => mclk, clr => clr, clk25 => clk25,
        clk190 => clk190);

U2 : debounce4
    port map(cclk => clk190, clr => clr, inp => btn,
        outp => btnd);

U3 : uart_tx
    port map(clk => clk25, clr => clr, tx_data => sw,
        ready => ready, tdre => tdre, TxD => TxD);

U4 : test_tx_ctrl
    port map(
        clk => clk25,
        clr => clr,
        go => btnd(0),
        tdre => tdre,
        ready => ready
    );

U5 : x7segb
    port map(x => x, cclk => clk190, clr => clr,
        a_to_g => a_to_g, an => an);

end uart_tx_top;
```

Example 34

Receive Module

Receiving data is similar to transmitting data. Fig. 5.7 shows the entity for the receiver. Eight bits of asynchronous data are input into *RxD*. Bits are shifted into the 8-bit shift register, *rx_data*(7:0), least-significant bit first. When *rx_data*(7:0) is full, the received-data ready flag, *rdrf*, is set to 1 to signify that *rx_data*(7:0) contains a complete byte. The output *rdrf* is set to 0 by setting *rdrf_clr* to high. The framing error flag, *FE*, is set to 1 if the stop bit is not 1. This is one indication that the data on *rx_data*(7:0) may not be accurate. Fig. 5.8 shows the state diagram for the serial receiver.

The state diagram in Fig. 5.8 has the same states as the state machine developed for transmitting data. The transitions between these states are also the same. There is, however, one subtle, important difference. Instead of remaining in the start state for a whole bit time, the state machine transitions to the delay state after a half-bit time. Bit time is determined by the baud rate in the same way it was used in the transmitter. Fig. 5.9 illustrates the difference in timing between serially transmitting and receiving 8 bits of data.

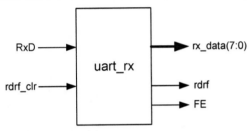

Figure 5.7 *uart_rx*

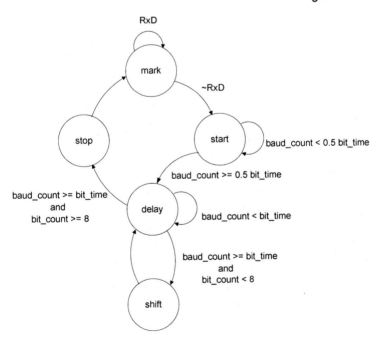

Figure 5.8 State diagram for receiving serial data

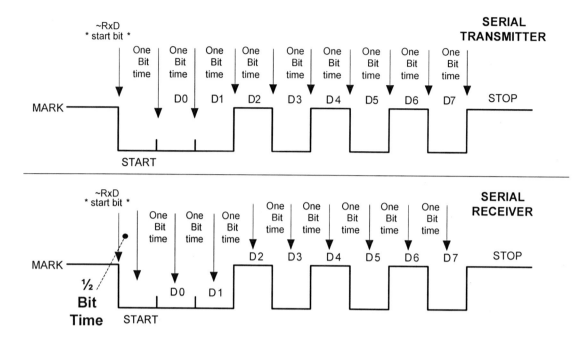

Figure 5.9 Timing differences between transmitting and receiving

During transmission, the start bit is transmitted for an entire bit time just like the data bits and the stop bit. When receiving, by waiting only a half bit time after the start bit has been initiated, the data shifted into the shift register during the shift state is farthest away from the time the signal changes. That is, half way between the beginning and ending time during which the bit is valid.

The state diagram in Fig. 5.8 starts in the *mark* state. This state resets a signal for counting the number of bits transmitted, *bit_count*, to zero and asserts the *rdrf* output low indicating that the data in *rx_data*(7:0) is not complete. When *RxD* goes low, signifying a start bit, the state diagram transitions to the *start* state. It remains in the *start* state for one-half of a bit time, then transitions to the *delay* state. The bit time counter, *baud_count* is incremented in the *delay* state and the state diagram remains in the *delay* state while the *baud_count* is less than the *bit_time*. After delaying for a bit time, the state diagram transitions to the *shift* state where a data bit is shifted into the shift register, *rx_data*(7:0). Additionally, *rdrf* remains low, *bit_count* is incremented to count the number of bits transmitted, and *baud_count* is reset to zero. Execution continues to the *delay* state for another bit time at which point the state diagram transitions to the *shift* state once again to store a bit of data. This continues until 8 bits of data have been shifted into *rx_data*(7:0). Finally, the state diagram transitions to the *stop* state where the *rdrf* flag is asserted high and *FE* is asserted high if the stop bit is not one as it should be. Execution ends in the *mark* state until *RxD* goes low once again.

Listing 5.4 shows the VHDL implementation of *uart_rx*.

Listing 5.4 uart_rx.vhd

```vhdl
-- Example 34a: UART receive component
library IEEE;
use IEEE.STD_LOGIC_1164.ALL;
use IEEE.STD_LOGIC_UNSIGNED.ALL;

entity uart_rx is
      port(
            RxD : in STD_LOGIC;
            clk : in STD_LOGIC;
            clr : in STD_LOGIC;
            rdrf_clr : in STD_LOGIC;
            rdrf : out STD_LOGIC;
            FE : out STD_LOGIC;
            rx_data : out STD_LOGIC_VECTOR(7 downto 0)
         );
end uart_rx;

architecture uart_rx of uart_rx is
type state_type is (mark, start, delay, shift, stop);
signal state: state_type;
signal rxbuff: STD_LOGIC_VECTOR (7 downto 0);
signal baud_count: STD_LOGIC_VECTOR (11 downto 0);
signal bit_count: STD_LOGIC_VECTOR (3 downto 0);
signal rdrf_set, fe_set, cclr, cclr8, rxload: STD_LOGIC;
constant bit_time: STD_LOGIC_VECTOR (11 downto 0) := X"A28";
constant half_bit_time: STD_LOGIC_VECTOR (11 downto 0) := X"514";

begin

uart2: process(clk, clr, rdrf_clr)
  begin
    if clr = '1' then
        state <= mark;
        rxbuff <= "00000000";
        baud_count <= X"000";
        bit_count <= "0000";
        rdrf <= '0';
        FE <= '0';
    elsif rdrf_clr = '1' then
        rdrf <= '0';
    elsif (clk'event and clk = '1') then
        case state is
           when mark =>              -- wait for start bit
              baud_count <= X"000";
              bit_count <= "0000";
              if RxD = '1' then
                    state <= mark;
              else
                    FE <= '0';
                    state <= start;        --   go to start
              end if;
```

Listing 5.4 (cont.) uart_rx.vhd

```vhdl
         when start =>                          -- check for start bit
             if baud_count >= half_bit_time then
                     baud_count <= X"000";
                     state <= delay;
             else
                     baud_count <= baud_count + 1;
                     state <= start;
             end if;

         when delay =>
             if baud_count >= bit_time then
                     baud_count <= X"000";
                     if bit_count < 8 then
                            state <= shift;
                     else
                            state <= stop;
                     end if;
             else
                     baud_count <= baud_count + 1;
                     state <= delay;
             end if;

         when shift =>                          -- get next bit
             rxbuff(7) <= RxD;
             rxbuff(6 downto 0) <= rxbuff(7 downto 1);
             bit_count <= bit_count + 1;
             state <= delay;
         when stop =>
             rdrf <= '1';
             if RxD = '0' then
                     FE <= '1';
             else
                     FE <= '0';
             end if;
             state <= mark;
      end case;
   end if;
end process uart2;

rx_data <= rxbuff;

end uart_rx;
```

Controlling *uart_rx*

Figure 5.10 shows a state diagram for *test_rx_ctrl* for testing the *uart_rx* component. The state diagram starts in the *wtrdrf* state and waits until the receive-data-ready flag, *rdrf*, goes high indicating that eight bits have been shifted into *rx_data*(7:0). The state diagram then transitions to the *load* state, which asserts *rdrf_clr* high to reset *rdrf* and signal the *uart_rx* to continue receiving data. Listing 5.5 shows the VHDL code for the *test_rx_ctrl* controller.

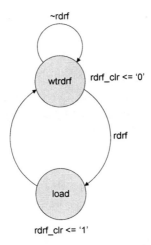

Figure 5.10 State diagram for *test_rx_ctrl*

Listing 5.5 test_rx_ctrl.vhd

```vhdl
-- Example 34b: Controller for Testing UART receiver
entity test_rx_ctrl is
    port(
            clk : in STD_LOGIC;
            clr : in STD_LOGIC;
            rdrf : in STD_LOGIC;
            rdrf_clr : out STD_LOGIC
        );
end test_rx_ctrl;

architecture test_rx_ctrl of test_rx_ctrl is
type state_type is (wtrdrf, load);
signal state: state_type;

    begin
    ctrl: process(clk, clr, rdrf)
      begin
        if clr = '1' then
                state <= wtrdrf;
                rdrf_clr <= '0';
        elsif (clk'event and clk = '1') then
           case state is
              when wtrdrf =>
                  rdrf_clr <= '0';
                  if rdrf = '0' then
                         state <= wtrdrf;
                  else
                         state <= load;
                  end if;
              when load =>
                  rdrf_clr <= '1';
                  state <= wtrdrf;
           end case;
        end if;
     end process ctrl;
   end test_rx_ctrl;
```

Echoing Serial Data

Fig. 5.11 shows a top-level design for echoing ASCII characters received over the serial port from a terminal program on a PC. The component *uart_rx* waits for a start bit and receives data until a byte has been received, asserting *rdrf* high. As described in Fig. 5.10, *test_rx_ctrl* transitions to the *load* state and asserts *rdrf_clr* high to clear the *rdrf* and then goes back to the *wtrdrf* state to wait for another byte to be shifted in. By connecting *rdrf_clr* to the *go* input of *text_tx_ctrl*, the component *uart_tx* is signaled to begin transmitting the received data byte out the *TxD* port. In this manner, a byte is received by the receiver and echoed back by immediately transmitting the byte. Data are echoed continuously. Listing 5.6 shows the VHDL code for this top-level design. A sample run of this program using *host.exe* is shown in Fig. 5.12.

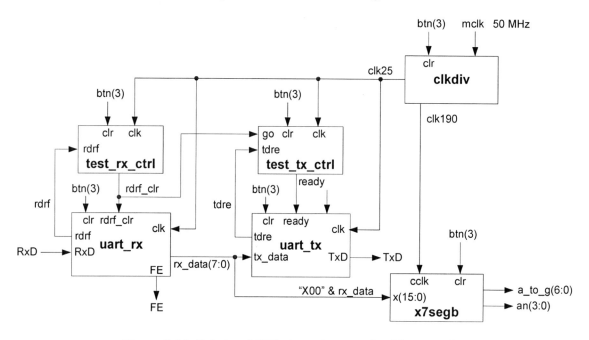

Figure 5.11 Echoing ASCII characters received from a PC

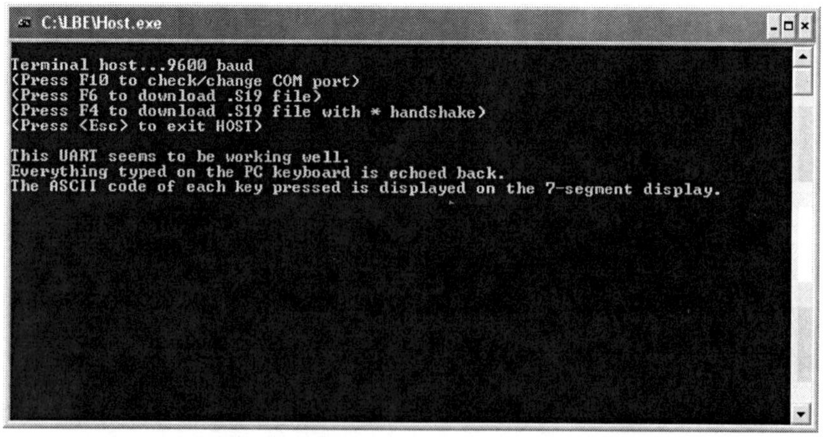

Figure 5.12 Sample run of Listing 5.6 using *host.exe*

Listing 5.6 uart_rx_top.vhd

```vhdl
-- Example 34c: uart_rx_top
library IEEE;
use IEEE.STD_LOGIC_1164.all;
use work.uart_components.all;

entity uart_rx_top is
    port(
            mclk : in STD_LOGIC;
            btn : in STD_LOGIC_VECTOR(3 downto 0);
            RxD : in STD_LOGIC;
            TxD : out STD_LOGIC;
            a_to_g : out STD_LOGIC_VECTOR(6 downto 0);
            an : out STD_LOGIC_VECTOR(3 downto 0)
        );
end uart_rx_top;

architecture uart_rx_top of uart_rx_top is
signal clk25, clk190, clr: std_logic;
signal tdre, ready, rdrf, rdrf_clr, FE: std_logic;
signal x: std_logic_vector(15 downto 0);
signal rx_data: std_logic_vector(7 downto 0);
signal btnd: std_logic_vector(3 downto 0);
begin
clr <= btn(3);
x <= X"00" & rx_data;
U1 : clkdiv
    port map(mclk => mclk, clr => clr, clk25 => clk25,
            clk190 => clk190);

U2 : debounce4
    port map(cclk => clk190, clr => clr, inp => btn,
            outp => btnd);

U3 : uart_tx
    port map(clk => clk25, clr => clr, tx_data => rx_data,
            ready => ready, tdre => tdre, TxD => TxD);

U4 : test_tx_ctrl
    port map(clk => clk25, clr => clr, go => rdrf_clr,
            tdre => tdre, ready => ready);

U5 : uart_rx
    port map(RxD => RxD, clk => clk25, clr => clr,
            rdrf_clr => rdrf_clr, rdrf => rdrf, FE => FE,
            rx_data => rx_data);

U6 : test_rx_ctrl
    port map(clk => clk25, clr => clr, rdrf => rdrf,
            rdrf_clr => rdrf_clr);

U7 : x7segb
    port map(x => x, cclk => clk190, clr => clr,
            a_to_g => a_to_g, an => an);

end uart_rx_top;
```

6

VGA Controller

A VGA controller is a component that controls five signals basic to video display. These signals are the horizontal sync *HS*, the vertical sync *VS*, and the three color signals *R*, *G*, and *B* which output the level of red, green, and blue, respectively. A VGA monitor works on the basis of emitting energy in the red, green, and blue spectrum proportional to the voltage on the corresponding *R*, *G*, or *B* signal input to the screen. Each colored dot on the screen is called a *pixel* (for picture element). How the monitor accomplishes this emission of color depends on the technology such as a cathode ray tube monitor or a liquid crystal display. In either case, the screen will display pixels starting from the top left corner of the screen moving towards the right and transitions, line-by-line in a direction moving towards the bottom of the screen. A horizontal sync pulse synchronizes each new line. Once it reaches the bottom of the screen, a vertical sync pulse causes it to start over again from the top left corner. The video controller must continuously output *R*, *G*, and *B* levels while synchronizing the horizontal lines and timing the vertical retrace to refresh the screen. The *R*, *G*, and *B* inputs to a monitor are analog. However, the ouput signals from the FPGA are digital outputs that need to be converted to an analog signal using some type of D/A converter. The Nexys-2 board uses a simple 3-resistor circuit of the type shown in Fig. 6.1 to convert a 3-bit red signal to *R*(2:0) to an 8-level analog signal V_R. It uses a similar circuit to convert three bits of green and two bits of blue to corresponding analog signals. Thus, the Nexys-2 board supports 8-bit VGA color – three bits of red, three bits of green, and two bits of blue. This will produce 256 different colors.

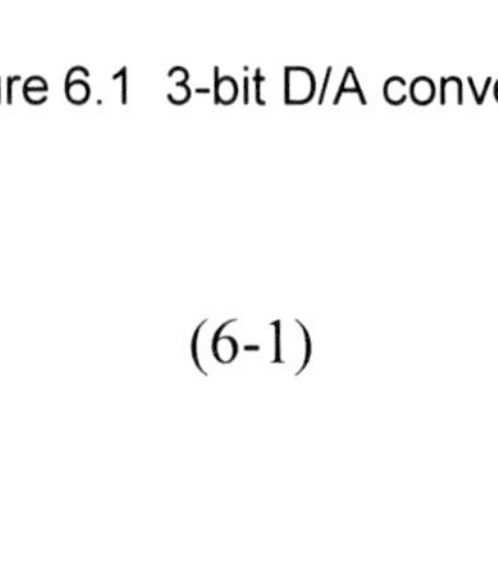

Figure 6.1 3-bit D/A converter

To analyze the circuit is Fig. 6.1 we can set the sum the currents at the node V_R to zero. Thus,

$$\frac{V_R - R_2}{0.5\text{k}\Omega} + \frac{V_R - R_1}{1\text{k}\Omega} + \frac{V_R - R_0}{2\text{k}\Omega} = 0 \tag{6-1}$$

Multiplying Eq. (6-1) by 2kΩ gives

$$4V_R - 4R_2 + 2V_R - 2R_1 + V_R - R_0 = 0$$

or

$$7V_R = 4R_2 + 2R_1 + R_0$$

from which

$$V_R = \frac{4}{7}R_2 + \frac{2}{7}R_1 + \frac{1}{7}R_0 \qquad (6\text{-}2)$$

Table 6.1 shows the eight possible values of V_R from Eq. (6-2) and these results are plotted in Fig. 6.2.

Table 6.1 Output of D/A converter

R_2	R_1	R_0	V_R
0	0	0	0
0	0	1	1/7
0	1	0	2/7
0	1	1	3/7
1	0	0	4/7
1	0	1	5/7
1	1	0	6/7
1	1	1	7/7

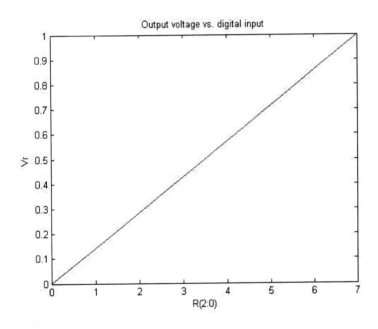

Figure 6.2 Plot of data in Table 6.1

Timing

A standard monitor operates at 60 Hz. That is, the screen is refreshed 60 times per second, which is equal to once every 1/60th of a second. Therefore, given a target resolution supported by the monitor, timing for each pixel, horizontal line, and vertical retrace can be derived.

We will design a VGA controller for a standard screen displaying a resolution of 640 x 480, a common, lower resolution for a VGA monitor. A 25 MHz clock will be used to drive the controller and as we will later discover, it is sufficient to accomplish the target resolution. We will refer to this as the pixel clock. First, we must derive the timing for the horizontal sync signal, *HS*. This signal consists of four regions, the sync pulse *(SP)*, back porch *(BP)*, horizontal video *(HV)*, and front porch *(FP)*. The sync pulse signals the beginning of a new line and is accomplished by bringing *HS* low. The signal is then brought high for the back porch where pixels are not yet written to the screen at the left. After the back porch, the *HS* signal remains high during the horizontal video period where pixels are written to the screen proceeding from left to right. Finally, the *HS* signal also remains high during the front porch where no pixels are written to the screen at the right. Fig. 6.3 shows the horizontal sync regions and timing.

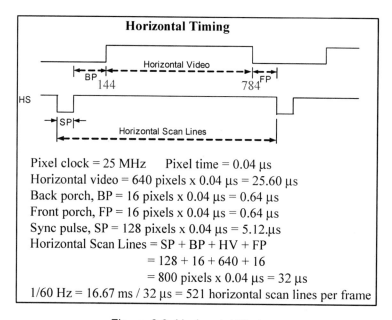

Figure 6.3 Horizontal Timing

At a 25 MHz pixel clock, each pixel will take $1/25 \times 10^6 = 0.04$ µs. Next, we should consider the horizontal video period since we desire 640 pixels per line for a 640 x 480 resolution. Six hundred forty pixels require 640 x 0.04 µs = 25.60 µs to display across one line. According to specification, the length of the sync pulse, *SP*, should be approximately one-fifth of the horizontal video time. Therefore, *SP* equals 25.60 µs / 5 = 5.12 µs. Using our 25 MHz clock, this means that *SP* requires 5.12µs / 0.04µs = 128 clock ticks or pixels. Also according to specification, the back and front porches should each be approximately one fortieth of the time required for the horizontal video. Therefore, the porches require 25.60 µs / 40 = 0.64 µs each. Using a 25 MHz pixel clock, each porch requires 0.64 µs / 0.04 µs = 16 clock ticks or pixels.

Finally, calculating the total number of clock pulses for a line we have *SP* + *BP* + *HV* + *FP* = 128 + 16 + 640 + 16 = 800 pixel clock pulses at 0.04 µs per pulse yields 32 µs for the entire line. Considering the line by counting pixels, we have 128 invisible pixels for the sync pulse where *HS* is low. The horizontal video region, where pixels are visible, starts at 128 + 16 = 144 after the sync pulse and invisible back porch and continues for 640 pixels stopping at 144 + 640 = 784. At this point, there is no visible video again at the front porch for 16 pixels until 784 + 16 = 800 where the sequence starts over again. Given that one entire screen, or frame, must be written in 1/60th of a second or 16.67 ms and each line requires 32 μs, it is possible to write 16.67 ms / 32 μs = 521 horizontal lines per screen. This is consistent with our target resolution of 640x480 which requires 480 lines per screen. Since 521 is slightly greater than 480, it is perfect for synchronizing the vertical sync signal since it also has a vertical sync pulse, back porch, vertical video, and front porch region.

Of the four regions that make up the vertical sync signal, the sync pulse *(SP)*, back porch *(BP)*, vertical video *(VV)*, and front porch *(FP)*, we already know that the vertical video region must be 480 lines. Fig. 6.4 shows the vertical sync signal and timing.

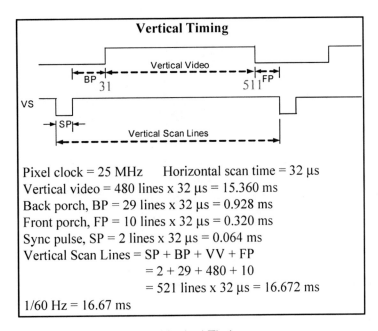

Figure 6.4 Vertical Timing

The vertical video region requires 480 lines at 32 μs per line = 15.360 ms. According to specification, the vertical sync pulse should be approximately 1/240th of the vertical video timing. Therefore, 15.360 ms / 240 = 0.064 ms for the vertical *SP*. At 32 μs per line, this requires 0.064 ms / 32 μs = 2 lines. Finally, splitting the remaining 39 lines between the back porch and the front porch using a 75%, 25% split as per specification, the back porch will be 29 lines and the front porch will be 10 lines. Finally, the 521 lines are split into a vertical SP + BP + VV + FP = 2 + 29 + 480 + 10. Considering the frame by counting lines, the *VS* signal is low for 2 lines to set the sync pulse *SP*. Then, *VS* is brought high for the remaining 519 lines. There is no visible video for the 29 line back porch followed by 480 lines of visible video until the counter has reached 31 + 480 = 511 where there is no visible video again for the 10 line front porch. The counter continues until it reaches 511 + 10 = 521 where the sequence starts over again and draws a new frame.

Other Resolutions

There are many other standard resolutions supported by most modern screens such as 800 x 600 and 1024 x 768. In order to achieve a higher resolution, a higher frequency pixel clock must be chosen. For example, to operate the VGA controller to output a resolution of 800 x 600, a 25 MHz clock would not be fast enough. Recall that with a 25 MHz clock and 640 visible pixels across each line, we only had 521 vertical lines. These lines had to be split between the vertical sync pulse, the vertical back and front porches, and the visible video. Obviously, if we expanded the pixels across to 800 visible pixels, the number of vertical lines would significantly decrease to much fewer than 521, while we desire 600 visible vertical lines. For any desired resolution, the pixel clock frequency must be calculated such that the following are true:

Given: a target HV and VV pixel resolution
Horizontal constraints
    HSP = 1/5 * HV   (horizontal sync pulse pixels)
    HBP = 1/40 * HV   (horizontal back porch pixels)
    HFP = 1/40 * HV   (horizontal front porch pixels)
Vertical constraints
    VSP = 1/240 * VV   (vertical sync pulse lines)
    VFP+VBP = 0.08125 * VV
    VBP = .75 * (VFP+VBP)   (vertical back porch lines)
    VFP = .25 * (VFP+VBP)   (vertical front porch lines)

where HSP+HBP+HV+HFP = HP_{total} pixels and VSP+VBP+VV+VFP = VP_{total} lines. Recall that VP_{total} must occur in 1/60th of a second. Then,
    16.67ms / VP_{total} = T_{line} ms
and T_{line} / HPtotal = T_{pixel} ms

Using these equations you can verify that a 64 MHz pixel clock would be required to achieve a 1024 x 768 resolution. In all of our examples we will use a 25 MHz clock to achieve a 640 x 480 resolution using the parameters in Figs. 6.3 and 6.4.

In this chapter we will illustrate how to write VHDL programs to design VGA controllers for the following examples:

        Example 35 – VGA-Stripes
        Example 36 – VGA-PROM
        Example 37 – Sprites in Block ROM
        Example 38 – Screen Saver
        Example 39 – External Video RAM
        Example 40 – External Video Flash

Example 35

VGA – Stripes

In this example we will write a VHDL program that will display fifteen green and red horizontal lines on the screen as shown in Fig. 6.5. We will find it convenient to design one component called *vga_640x480* that will generate the horizontal and vertical sync pulses, *hsync* and *vsync*, and a second component called *vga_stripes* that will generate the *red*, *green*, and *blue* outputs as shown in the top-level design in Fig. 6.6. We will be able to use the *vga_640x480* component in all of our examples in this chapter. We then only need to change the *vga_stripes* component to generate different color displays.

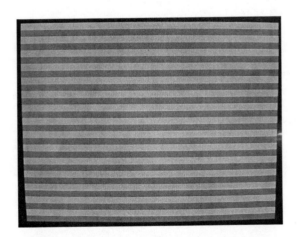
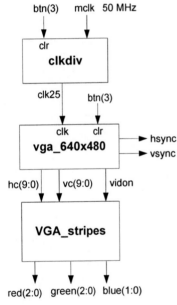

Figure 6.5 Screen display for Example 35   Figure 6.6 Top-level design for Example 35

Listings 6.1 shows the VHDL implementation of the *vga_640x480* component in Fig. 6.6. As input, the entity defines a 25 MHz clock, *clk*, and a clear signal, *clr*. We will create a horizontal counter that counts pixels and a vertical counter to count lines. The controller outputs *hsync*, *vsync*, horizontal and vertical counters, *hc(9:0)* and *vc(9:0)*, and *vidon* that is 1 when the horizontal and vertical counters are within the 640 x 480 display area.

Listing 6.1 vga_640x480.vhd

```vhdl
-- Example 35a: vga_640x480
library IEEE;
use IEEE.STD_LOGIC_1164.ALL;
use IEEE.STD_LOGIC_UNSIGNED.ALL;

entity vga_640x480 is
    port ( clk, clr : in std_logic;
           hsync : out std_logic;
           vsync : out std_logic;
           hc : out std_logic_vector(9 downto 0);
           vc : out std_logic_vector(9 downto 0);
           vidon : out std_logic
    );
end vga_640x480;

architecture vga_640x480 of vga_640x480 is
constant hpixels: std_logic_vector(9 downto 0) := "1100100000";
    --Value of pixels in a horizontal line = 800
constant vlines: std_logic_vector(9 downto 0) := "1000001001";
    --Number of horizontal lines in the display = 521
constant hbp: std_logic_vector(9 downto 0) := "0010010000";
    --Horizontal back porch = 144 (128+16)
constant hfp: std_logic_vector(9 downto 0) := "1100010000";
    --Horizontal front porch = 784 (128+16+640)
constant vbp: std_logic_vector(9 downto 0) := "0000011111";
    --Vertical back porch = 31 (2+29)
constant vfp: std_logic_vector(9 downto 0) := "0111111111";
    --Vertical front porch = 511 (2+29+480)
signal hcs, vcs: std_logic_vector(9 downto 0);
    --These are the Horizontal and Vertical counters
signal vsenable: std_logic;
    --Enable for the Vertical counter

begin
    --Counter for the horizontal sync signal
    process(clk, clr)
    begin
       if clr = '1' then
          hcs <= "0000000000";
       elsif(clk'event and clk = '1') then
          if hcs = hpixels - 1 then

             --The counter has reached the end of pixel count
                hcs <= "0000000000";
                --reset the counter
                vsenable <= '1';   --Enable the vertical counter
             else
                hcs <= hcs + 1;
                --Increment the horizontal counter
                vsenable <= '0';
                --Leave the vsenable off
          end if;
       end if;
    end process;
```

Listing 6.1 (cont.) vga_640x480.vhd

```vhdl
        hsync <= '0' when hcs < 128 else '1';
           --Horizontal Sync Pulse is low when hc is 0 - 127

    --Counter for the vertical sync signal
    process(clk, clr)
    begin
       if clr = '1' then
           vcs <= "0000000000";
       elsif(clk'event and clk = '1' and vsenable='1') then
           --Increment when enabled
           if vcs = vlines - 1 then
                --Reset when the number of lines is reached
                vcs <= "0000000000";
           else
                vcs <= vcs + 1;    --Increment vertical counter
           end if;
       end if;
    end process;

    --Vertical Sync Pulse is low when vc is 0 or 1
    vsync <= '0' when vcs < 2 else '1';

    --Enable video out when within the porches
    vidon <= '1' when (((hcs < hfp) and (hcs >= hbp))
                  and ((vcs < vfp) and (vcs >= vbp))) else '0';

    -- output horizontal and vertical counters
    hc <= hcs;
    vc <= vcs;

end vga_640x480;
```

For convenience, we define constants for the horizontal and vertical regions including the horizontal and vertical sync pulses, back porches, video regions, and front porches. The constants will define the regions according to the horizontal and vertical counter values.

Listing 6.1 shows a process for the horizontal sync counter. This counter must count up to 800 decimal and therefore must be 10 bits wide. When the counter equals *hpixels*, it has reached 800 and must start over. At that time, it also must signal the vertical line counter to count a line by bringing *vsenable* high for one clock cycle. Finally, the *hsync* signal is low during the sync pulse and back porch, that is, when *hcs* is less than 128. Otherwise, *hsync* is high. At this point, we are only setting the *hsync* signal. We will consider another signal, *vidon*, to differentiate the visible video ranges from the porches.

Listing 6.1 shows a similar process for the vertical sync counter. Every time the *vsenable* signal goes high, the vertical counter increments to count a line. When the counter has reached *vlines*, it must reset to start counting lines for a new frame. Finally, *vsync* is set as low if *vc* is less than 2 for the vertical sync pulse and high otherwise.

Now that we have implemented a counter for the horizontal sync and a counter for the vertical sync signals and have defined *hsync* and *vsync* accordingly, we must differentiate between regions with visible video from the porches. In Listing 6.1, we set the *vidon* signal to high when the horizontal counter is between the horizontal back porch and front porch and the vertical counter is between the vertical back porch and front porch. Otherwise, *vidon* is low where there is no visible video.

Listing 6.2 shows the VHDL implementation of the *vga_stripes* component in Fig. 6.6. A process for driving the *red, green,* and *blue* signals is given. All signals are defaulted to low. In the visible video regions, when *vidon* is high, all three *red* outputs are high when *vc*(4) is high and all three *green* outputs are low when *vc*(4) is high. This way, every sixteen lines the colors will alternate between red and green showing a band of green sixteen lines high and a band of red sixteen lines high.

Listing 6.3 shows the VHDL program for the top-level design in Fig. 6.6. A simulation for 17.0 ms, which is slightly longer than the time required to display one frame, is shown in Fig. 6.7. Notice the *vsync* signal pulse in the beginning and at the 16.67 ms time indicating the start of a new frame. Also note the back and front porches shown when *vidon* is low. Finally, note that the red outputs are high and the green outputs are low when *vc*(4) is high. Fig. 6.8 shows a simulation for only the first 36 visible lines which alternate between red and green every sixteen lines.

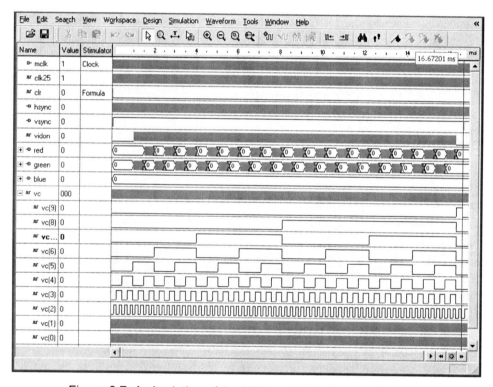

Figure 6.7 A simulation of the VGA controller for one frame

VGA – Stripes

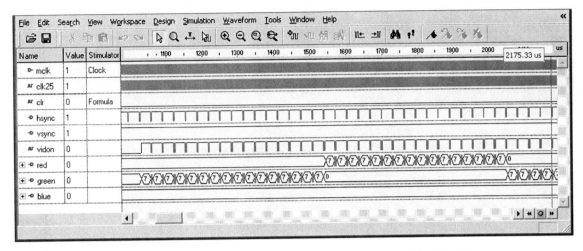

Figure 6.8 A simulation of the VGA controller for the first 36 visible lines

Listing 6.2 vga_stripes.vhd

```vhdl
-- Example 35b: vga_stripes
library IEEE;
use IEEE.STD_LOGIC_1164.ALL;
use IEEE.STD_LOGIC_UNSIGNED.ALL;

entity vga_stripes is
    port ( vidon: in std_logic;
           hc : in std_logic_vector(9 downto 0);
           vc : in std_logic_vector(9 downto 0);
           red : out std_logic_vector(2 downto 0);
           green : out std_logic_vector(2 downto 0);
           blue : out std_logic_vector(1 downto 0));
end vga_stripes;

architecture vga_stripes of vga_stripes is
begin

    process(vidon, vc)
    begin
          red <= "000";
          green <= "000";
          blue <= "00";
          if vidon = '1' then
                red <= vc(4) & vc(4) & vc(4);
                green <= not (vc(4) & vc(4) & vc(4));
          end if;
    end process;

end vga_stripes;
```

Listing 6.3 vga_stripes_top.vhd

```vhdl
library IEEE;
use IEEE.STD_LOGIC_1164.all;
use work.vga_components.all;

entity vga_stripes_top is
    port(
          mclk : in STD_LOGIC;
          btn : in STD_LOGIC_VECTOR(3 downto 0);
          hsync : out STD_LOGIC;
          vsync : out STD_LOGIC;
          red : out std_logic_vector(2 downto 0);
          green : out std_logic_vector(2 downto 0);
          blue : out std_logic_vector(1 downto 0)
         );
end vga_stripes_top;

architecture vga_stripes_top of vga_stripes_top is
signal clr, clk25, vidon: std_logic;
signal hc, vc: std_logic_vector(9 downto 0);
begin
    clr <= btn(3);

U1 : clkdiv
    port map(
          mclk => mclk,
          clr => clr,
          clk25 => clk25
    );

U2 : vga_640x480
    port map(
          clk => clk25,
          clr => clr,
          hsync => hsync,
          vsync => vsync,
          hc => hc,
          vc => vc,
          vidon => vidon
    );

U3 : vga_stripes
    port map(
          vidon => vidon,
          hc => hc,
          vc => vc,
          red => red,
          green => green,
          blue => blue
    );

end vga_stripes_top;
```

Example 36

VGA – PROM

In this example we will use a ROM of the type we introduced in Listing 4.1 of Example 27 to design a VGA controller to display your initials on the screen. Listing 6.4 shows a ROM containing the initials DMH. In a sense, it is a type of programmable read-only memory since we can change its contents in VHDL and reprogram our FPGA. Therefore, the entity has been named *prom_DMH*.

Listing 6.4 prom_DMH.vhd

```vhdl
-- Example 36a: prom_DMH
library IEEE;
use IEEE.std_logic_1164.all;
use IEEE.std_logic_unsigned.all;

entity prom_DMH is
    port (
        addr: in STD_LOGIC_VECTOR (3 downto 0);
        M: out STD_LOGIC_VECTOR (0 to 31)
    );
end prom_DMH;

architecture prom_DMH of prom_DMH is
type rom_array is array (NATURAL range <>)
              of STD_LOGIC_VECTOR (0 to 31);
constant rom: rom_array := (
    "01111110000011000001101000000010",   --0
    "01000001000011000001101000000010",   --1
    "01000000100010100010101000000010",   --2
    "01000000010010100010101000000010",   --3
    "01000000001010100010101000000010",   --4
    "01000000001010010100101000000010",   --5
    "01000000001010010100101000000010",   --6
    "01000000001010010100101111111110",   --7
    "01000000001010001000101000000010",   --8
    "01000000001010001000101000000010",   --9
    "01000000001010001000101000000010",   --10
    "01000000001010000000101000000010",   --11
    "01000000010010000000101000000010",   --12
    "01000000100010000000101000000010",   --13
    "01000001000010000000101000000010",   --14
    "01111110000010000000101000000010"    --15
    );
begin
  process(addr)
    variable j: integer;
    begin
      j := conv_integer(addr);
      M <= rom(j);
   end process;
end prom_DMH;
```

The PROM shown in Listing 6.4 contains sixteen data entries, each a 32-bit word. One use of this PROM would be to indicate whether or not a pixel should be drawn on the screen. A zero would indicate that the pixel should be turned off, while a one indicates that a pixel should be appropriately placed on the screen. Notice that the ones in this PROM spell the initials DMH among zeros that make up the background. This 16 by 32 image contained in the PROM can be used in conjunction with the *vga_640x480* component from Example 35 to display those initials on the screen. This 2-dimensional smaller image integrated into a larger scene is called a *sprite*.

A top-level design to display the initials on the screen is shown in Fig. 6.9. Listing 6.5 shows the VHDL program for the *vga_initials* component. The switch input *sw*(7:0) will control the location of the initials on the screen.

In Listing 6.5 the constants *w* and *h* refer to the width and height respectively of the sprite containing the initials. In this case, the initials have been created using 32 bits across and 16 lines high according to the data in the PROM. The signals *C1* and *R1* keep track of the current column and row of the upper-left position of the sprite. Signals *rom_addr4*(3:0) and *M*(0:31) are used when wiring the *prom_DMH* component to the *vga_initials* component. While the *rom_addr4* signal will refer to the current address for the PROM, the *rom_pix* signal will refer to a particular bit in the word that is being addressed. In this fashion, *rom_addr4* is related to the row and *rom_pix* is related to the column. Together, they refer to a single bit in the bitmap which will indicate whether or not a pixel should be placed at the corresponding location on the screen.

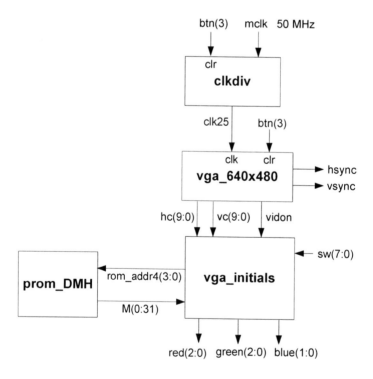

Figure 6.9 Top-level design for Example 36

Listing 6.5 vga_initials.vhd

```vhdl
-- Example 36b: vga_initials
library IEEE;
use IEEE.STD_LOGIC_1164.ALL;
use IEEE.STD_LOGIC_UNSIGNED.ALL;

entity vga_initials is
    port ( vidon: in std_logic;
          hc : in std_logic_vector(9 downto 0);
          vc : in std_logic_vector(9 downto 0);
          M: in std_logic_vector(0 to 31);
          sw: in std_logic_vector(7 downto 0);
          rom_addr4: out std_logic_vector(3 downto 0);
          red : out std_logic_vector(2 downto 0);
          green : out std_logic_vector(2 downto 0);
          blue : out std_logic_vector(1 downto 0)
        );
end vga_initials;

architecture vga_initials of vga_initials is
constant hbp: std_logic_vector(9 downto 0) := "0010010000";
constant vbp: std_logic_vector(9 downto 0) := "0000011111";
constant w: integer := 32;
constant h: integer := 16;
signal C1, R1: std_logic_vector(10 downto 0);
signal rom_addr, rom_pix: std_logic_vector(10 downto 0);
signal spriteon, R, G, B: std_logic;
begin
    --set C1 and R1 using switches
    C1 <= "00" & SW(3 downto 0) & "00001";
    R1 <= "00" & SW(7 downto 4) & "00001";
    rom_addr <= vc - vbp - R1;
    rom_pix <= hc - hbp - C1;
    rom_addr4 <= rom_addr(3 downto 0);
    --Enable sprite video out when within the sprite region
    spriteon <= '1' when (((hc >= C1 + hbp) and (hc < C1 + hbp + w))
          and ((vc >= R1 + vbp) and (vc < R1 + vbp + h))) else '0';

    process(spriteon, vidon, rom_pix, M)
    variable j: integer;
    begin
        red <= "000";
        green <= "000";
        blue <= "00";
        if spriteon = '1' and vidon = '1' then
            j := conv_integer(rom_pix);
            R <= M(j);
            G <= M(j);
            B <= M(j);
            red <= R & R & R;
            green <= G & G & G;
            blue <= B & B;
        end if;
    end process;

end vga_initials;
```

The signal *spriteon* is used to indicate whether or not the current pixel given by (*vc, hc*) is in the region where the sprite is to be drawn. This *spriteon* signal is similar to the *vidon* signal in the *vga_640x480* component in that it is high if the current pixel coordinate designated by (*vc, hc*) is in the 16 x 32 region designated for the sprite. To accomplish this, it is defined to be high as a function of both *hc* and *vc*. For it to be high, *hc* must be between *hbp* + *C*1 and *hbp*+*C*1+*w*. That is, between the beginning of the visible horizontal range plus the column offset, *C*1, plus the width of the sprite, *w*, which is 32 in this example. Similarly, *vc* must be between *vbp* + *R*1 plus the sprite height, *h*, which is 16 in this example. These conditions define the visible sprite area.

*C*1 and *R*1 are set by the switches as shown. At a minimum, the sprite can start at pixel coordinate (1,1) and at a maximum (481, 481) which is off the screen. The first line of the sprite region corresponds to the first row in the sprite bitmap which is at address zero. Likewise, the first column of the sprite region corresponds to the first pixel in the row of data, pixel zero. Therefore, *rom_addr*(9:0) is set to *vc* – *vbp* – *R*1 and *rom_pix* is set to *hc* – *hbp* – *C*1. The output *rom_addr4*(3:0) consists of the lower four bits of *rom_addr*(9:0). Together, the coordinate (*rom_addr4, rom_pix*) identifies the pixel in the sprite bitmap.

The process in Listing 6.5 controls the pixel colors where the initials are displayed white in this example. The pixels are only displayed in the visible video region (*vidon* = '1') and visible sprite region (*spriteon* = '1'). Whether they are on or off in that region depends on the jth bit of *M*(0:31) where *j* is the integer value of *rom_pix* (the column) and *M*(0:31) is the data from the PROM at address *rom_addr4*(3:0).

Fig. 6.10 illustrates how the *rom_addr* and *rom_pix* are calculated as the *vc* and *hc* proceed to increment relative to the *vbp, hbp*, and selected location for the sprite, *R*1 and *C*1. The top of Fig. 6.10 shows the display screen. The bitmap for the intials in the sprite PROM is shown at the bottom of the figure. The example in Fig. 6.10 uses a selected sprite position of *R*1 = 1 and *C*1 = 1. The *rom_addr* and *rom_pix* will therefore be calculated as shown in the legend labeled '*GIVEN*' in Fig 6.10. As the vertical counter, *vc*, and horizontal counter, *hc*, increment past the horizontal and vertical blanking periods, to the first pixel where video is visible (32,145), *spriteon* is high. As shown in *Sample* 1, the *rom_addr* = 0 yielding the first row of data in the sprite bitmap from the PROM while *rom_pix* = 0 selecting the first bit of that data word. As the horizontal counter continues across the screen, *rom_pix* also moves across the data word. Once the horizontal counter increments past the visible sprite region to *hc* = 177, *spriteon* becomes low and data from the PROM is not relevant. Finally, as the vertical counter moves down the screen, *rom_addr* moves to successive data words. Similarly, once the vertical counter has incremented to *vc* = 48, *spriteon* becomes low and data from the PROM is no longer relevant.

Listing 6.6 in the VHDL program for the top-level design shown in Fig. 6.9. When you run this program the initials DMH will be displayed on the video screen at a location determined by the switch settings. Modify the PROM contents in Listing 6.4 to display your own initials. Download and run your resulting program.

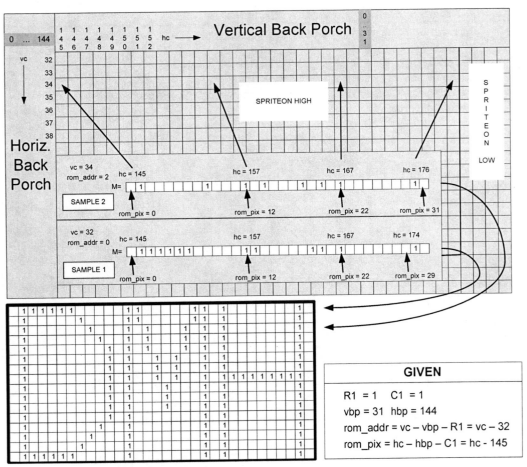

Figure 6.10 Drawing the sprite on the screen

Listing 6.6 vga_initials_top.vhd

```
-- Example 36c: vga_initials_top
library IEEE;
use IEEE.STD_LOGIC_1164.all;
use work.vga_components.all;

entity vga_initials_top is
        port(
           mclk : in STD_LOGIC;
           btn : in STD_LOGIC_VECTOR(3 downto 0);
           sw : in STD_LOGIC_VECTOR(7 downto 0);
           hsync : out STD_LOGIC;
           vsync : out STD_LOGIC;
           red : out std_logic_vector(2 downto 0);
           green : out std_logic_vector(2 downto 0);
           blue : out std_logic_vector(1 downto 0)
             );
end vga_initials_top;
```

Listing 6.6 (cont.) vga_initials_top.vhd

```vhdl
architecture vga_initials_top of vga_initials_top is
signal clr, clk25, vidon: std_logic;
signal hc, vc: std_logic_vector(9 downto 0);
signal M: std_logic_vector(0 to 31);
signal rom_addr4: std_logic_vector(3 downto 0);
begin

      clr <= btn(3);

U1 : clkdiv
     port map(mclk => mclk, clr => clr, clk25 => clk25);

U2 : vga_640x480
     port map(clk => clk25, clr => clr, hsync => hsync,
              vsync => vsync, hc => hc, vc => vc,
              vidon => vidon);

U3 : vga_initials
     port map(vidon => vidon, hc => hc, vc => vc, M => M,
              sw => sw, rom_addr4 => rom_addr4, red => red,
              green => green, blue => blue);

U4 : prom_dmh
     port map(addr => rom_addr4, M => M);

end vga_initials_top;
```

Example 37

Sprites in Block ROM

In Example 36 we made a sprite by storing the bit map of three initials in a VHDL ROM. To make a larger sprite we could use the Core Generator to store the sprite image in either distributed or block RAM/ROM. Block ROM will allow larger images to be stored and in this example we will use a block ROM to store the image of the two loons shown in Fig. 6.11. The size of this image is 240 x 160 pixels and is available at www.lbebooks.com as the image file *loons240x160.jpg*.

In Example 30 we showed how to create a block ROM using the Core Generator where the initial data is stored in a *.coe* file. We can convert the JPEG image in the file *loons240x160.jpg* to a corresponding *.coe* file called *loons240x160.coe* by using the Matlab function *IMG2coe8(imgfile, outfile)* shown in Listing 6.7. This function will work with other standard image formats other than JPEG such as *bmp*, *gif*, and *tif*. Note that the *.coe* file produced by this function contains an 8-bit byte for each image pixel with the format

$$\text{color byte} = [R2,R1,R0,G2,G1,G0,B1,B0] \qquad (6\text{-}3)$$

The original image read into the Matlab function in Listing 6.7 will contain 8-bits of red, 8-bits of green, and 8-bits of blue. The 8-bit color byte stored in the *.coe* file will contain only the upper 3 bits of red, the upper 3 bits of green, and the upper 2 bits of blue. We need to do this because as we have seen the Nexys-2 board supports only 8-bit VGA colors. The resulting 8-bit color image is called *img2* in Listing 6.7 and will be of reduced quality from the original image as can be seen in Fig. 6.12.

Figure 6.11 Loon photo by Edie Haskell

Figure 6.12
The image *img2* produced by Listing 6.7

Listing 6.7 IMG2coe8.m

```matlab
function img2 = IMG2coe8(imgfile, outfile)
% Create .coe file from .jpg image
% .coe file contains 8-bit words (bytes)
% each byte contains one 8-bit pixel
% color byte: [R2,R1,R0,G2,G1,G0,B1,B0]
% img2 = IMG2coe8(imgfile, outfile)
% img2 is 8-bit color image
% imgfile = input .jpg file
% outfile = output .coe file
% Example:
% img2 = IMG2coe8('loons240x160.jpg', 'loons240x160.coe');

img = imread(imgfile);
height = size(img, 1);
width = size(img, 2);
s = fopen(outfile,'wb');   %opens the output file
fprintf(s,'%s\n',';  VGA Memory Map ');
fprintf(s,'%s\n','; .COE file with hex coefficients ');
fprintf(s,'; Height: %d, Width: %d\n\n', height, width);
fprintf(s,'%s\n','memory_initialization_radix=16;');
fprintf(s,'%s\n','memory_initialization_vector=');
cnt = 0;
img2 = img;
for r=1:height
    for c=1:width
        cnt = cnt + 1;
        R = img(r,c,1);
        G = img(r,c,2);
        B = img(r,c,3);
        Rb = dec2bin(R,8);
        Gb = dec2bin(G,8);
        Bb = dec2bin(B,8);
        img2(r,c,1) = bin2dec([Rb(1:3) '00000']);
        img2(r,c,2) = bin2dec([Gb(1:3) '00000']);
        img2(r,c,3) = bin2dec([Bb(1:2) '000000']);
        Outbyte = [ Rb(1:3) Gb(1:3) Bb(1:2) ];
        if (Outbyte(1:4) == '0000')
            fprintf(s,'0%X',bin2dec(Outbyte));
        else
            fprintf(s,'%X',bin2dec(Outbyte));
        end
        if ((c == width) && (r == height))
            fprintf(s,'%c',';');
        else
            if (mod(cnt,32) == 0)
                fprintf(s,'%c\n',',');
            else
                fprintf(s,'%c',',');
            end
        end
    end
end
fclose(s);
```

The *.coe* file produced by Listing 6.7 will contain $240 \times 160 = 38,400$ bytes. Following the procedure described in Example 30 we can use the Core Generator to create a read only block memory component called *loons240x160*, which is 8 bits wide with a depth of 38400 as shown in Fig. 6.13. This component would be part of the top-level design shown in Fig. 6.14.

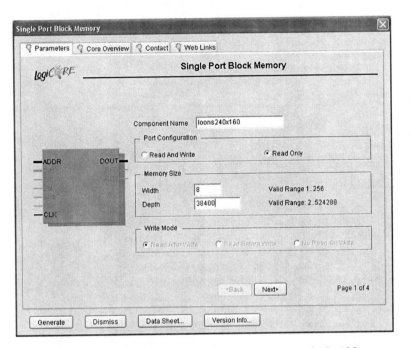

Figure 6.13 Generating the component *loons240x160*

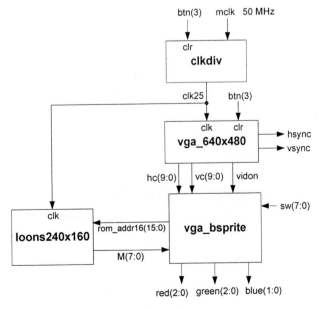

Figure 6.14 Top-level design to display the loon photo on the VGA screen

Listing 6.8 is the VHDL program for the component *vga_bsprite* shown in Fig. 6.14. It is similar to the component *vga_initials* shown in Fig. 6.9 of Example 36 and described by Listing 6.5. The main difference is that *rom_addr16*(15:0) is now calculated from the formula

$$rom\_addr16 = ypix \times 240 + xpix$$

where *xpix* and *ypix* are the local image coordinates within the loon image of the pixel whose color byte is at address *rom_addr16* in the block ROM *loons240x160*.

Listing 6.8 vga_bsprite.vhd

```vhdl
-- Example 37a: vga_bsprite
library IEEE;
use IEEE.STD_LOGIC_1164.ALL;
use IEEE.STD_LOGIC_UNSIGNED.ALL;

entity vga_bsprite is
    port ( vidon: in std_logic;
           hc : in std_logic_vector(9 downto 0);
           vc : in std_logic_vector(9 downto 0);
           M: in std_logic_vector(7 downto 0);
           sw: in std_logic_vector(7 downto 0);
           rom_addr16: out std_logic_vector(15 downto 0);
           red : out std_logic_vector(2 downto 0);
           green : out std_logic_vector(2 downto 0);
           blue : out std_logic_vector(1 downto 0)
         );
end vga_bsprite;

architecture vga_bsprite of vga_bsprite is
constant hbp: std_logic_vector(9 downto 0) := "0010010000";
    --Horizontal back porch = 144 (128+16)
constant vbp: std_logic_vector(9 downto 0) := "0000011111";
    --Vertical back porch = 31 (2+29)
constant w: integer := 240;
constant h: integer := 160;
signal xpix, ypix: std_logic_vector(9 downto 0);
signal C1, R1: std_logic_vector(9 downto 0);
signal spriteon, R, G, B: std_logic;

begin
    --set C1 and R1 using switches
    C1 <= '0' & SW(3 downto 0) & "00001";
    R1 <= '0' & SW(7 downto 4) & "00001";
    ypix <= vc - vbp - R1;
    xpix <= hc - hbp - C1;

    --Enable sprite video out when within the sprite region
    spriteon <= '1' when (((hc >= C1 + hbp) and (hc < C1 + hbp + w))
        and ((vc >= R1 + vbp) and (vc < R1 + vbp + h))) else '0';
```

Listing 6.8 (cont.) vga_bsprite.vhd

```vhdl
        process(xpix, ypix)
        variable  rom_addr1, rom_addr2: STD_LOGIC_VECTOR (16 downto 0);
        begin
                rom_addr1 := (ypix & "0000000") + ('0' & ypix & "000000")
                    + ("00" & ypix & "00000") + ("000" & ypix & "0000");
                    -- y*(128+64+32+16) = y*240
                rom_addr2 := rom_addr1 + ("00000000" & xpix);   -- y*240+x
                rom_addr16 <= rom_addr2(15 downto 0);
        end process;

        process(spriteon, vidon, M)
              variable j: integer;
        begin
                red <= "000";
                green <= "000";
                blue <= "00";
                if spriteon = '1' and vidon = '1' then
                      red <= M(7 downto 5);
                      green <= M(4 downto 2);
                      blue <= M(1 downto 0);
                end if;
        end process;

end vga_bsprite;
```

Notice how we multiply *ypix* by 240 using a shift and add method by recognizing that

$$ypix \times 240 = ypix \times (128 + 64 + 32 + 16)$$

The *red*, *green*, and *blue* outputs are obtained from the *loon240x160* ROM output $M(7:0)$ according to the color byte format shown in Eq. (6-3).

Listing 6.9 is the VHDL program for the top-level design shown in Fig. 6.14.

Listing 6.9 vga_bsprite_top.vhd

```vhdl
-- Example 37b: vga_bsprite_top
library IEEE;
use IEEE.STD_LOGIC_1164.all;
use work.vga_components.all;

entity vga_bsprite_top is
        port(
                mclk : in STD_LOGIC;
                btn : in STD_LOGIC_VECTOR(3 downto 0);
                sw : in STD_LOGIC_VECTOR(7 downto 0);
                hsync : out STD_LOGIC;
                vsync : out STD_LOGIC;
                red : out std_logic_vector(2 downto 0);
                green : out std_logic_vector(2 downto 0);
                blue : out std_logic_vector(1 downto 0)
             );
end vga_bsprite_top;

architecture vga_bsprite_top of vga_bsprite_top is
signal clr, clk25, vidon: std_logic;
signal hc, vc: std_logic_vector(9 downto 0);
signal M: std_logic_vector(7 downto 0);
signal rom_addr16: std_logic_vector(15 downto 0);
begin

     clr <= btn(3);

U1 : clkdiv
     port map(mclk => mclk, clr => clr, clk25 => clk25);

U2 : vga_640x480
     port map(clk => clk25, clr => clr, hsync => hsync,
     vsync => vsync, hc => hc, vc => vc, vidon => vidon);

U3 : vga_bsprite
     port map(vidon => vidon, hc => hc, vc => vc, M => M, sw => sw,
          rom_addr16 => rom_addr16, red => red, green => green,
          blue => blue);

U4 : loons240x160
     port map (addr => rom_addr16, clk => clk25, dout => M);

end vga_bsprite_top;
```

Example 38

Screen Saver

In this example we will modify Example 37 by putting the loon picture in motion to form a kind of screen saver in which the photo will bounce off the edges of the screen at 45° angles as shown in Fig. 6.15. The upper-left corner of the photo is located at column $C1$ and row $R1$. When you start the program or press $btn(3)$ we will display the loon photo at the initial location $C1 = 80$, $R1 = 140$ shown in Fig. 6.15. Pressing $btn(0)$ will set the photo in motion along the path shown in Fig. 6.15.

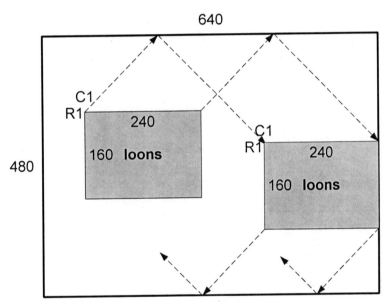

Figure 6.15 Motion of loon photo as a screen saver

It is clear from Fig. 6.15 that to move the photo all we need to do is to change the values of $C1$ and $R1$. These values were controlled by the switches in Example 37, but in this example we will control the values of $C1$ and $R1$ using a separate component called *bounce* as shown in the top-level design in Fig. 6.16. Thus, the input $sw(7:0)$ to *vga_bsprite* in Listing 6.8 is replaced by the two inputs $C1(9:0)$ and $R1(9:0)$ in the *vga_ScreenSaver* component whose VHDL program is shown in Listing 6.10.

The component *bounce* in Fig. 6.16 will change the values of $C1$ and $R1$ according to the algorithm shown in Fig. 6.17. Starting at the initial location $C1 = 80$, $R1 = 140$, the value of $C1$ is incremented by $\Delta C1$ and the value of $R1$ is decremented by $\Delta R1$. This will cause the photo to move upward to the right. Choosing the magnitude of $\Delta C1$ and $\Delta R1$ to be 1 will produce a smooth motion of the photo. The speed of the moving photo will be determined by how fast the instructions in the algorithm in Fig. 6.17 are executed.

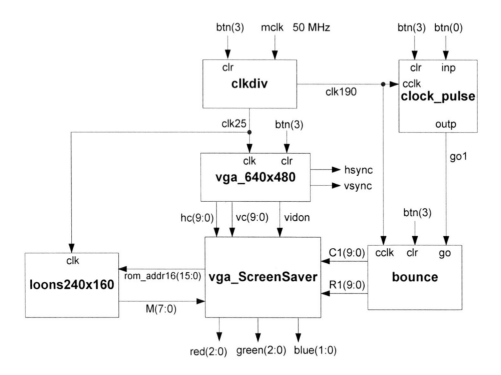

Figure 6.16 Top-level design of screen saver

bounce:
    $C1 = 80;$
    $R1 = 140;$
    $\Delta C1 = 1;$
    $\Delta R1 = -1;$
    while(1){
        $C1 = C1 + \Delta C1;$
        $R1 = R1 + \Delta R1;$
        if $(C1 < 0 \text{ or } C1 \geq C1\max)$
            $\Delta C1 = -\Delta C1;$
        if $(R1 < 0 \text{ or } R1 \geq R1\max)$
            $\Delta R1 = -\Delta R1;$
    }

Figure 6.17 Algorithm for the component *bounce*

The photo will reach the top of the screen when $R1$ becomes zero and will reach the bottom of the screen when $R1$ becomes $R1\max = 480 - 160 = 320$. The photo will reach the left edge of the screen when $C1$ becomes zero and will reach the right edge of the screen when $C1$ becomes $C1\max = 640 - 240 = 400$. According to the algorithm in Fig. 6.17 when the photo reaches the edge of the screen the sign of $\Delta C1$ or $\Delta R1$ is reversed so that the photo will reverse direction.

Listing 6.10 vga_ScreenSaver.vhd

```vhdl
-- Example 38a: vga_ScreenSaver
library IEEE;
use IEEE.STD_LOGIC_1164.ALL;
use IEEE.STD_LOGIC_UNSIGNED.ALL;

entity vga_ScreenSaver is
    port ( vidon: in std_logic;
           hc : in std_logic_vector(9 downto 0);
           vc : in std_logic_vector(9 downto 0);
           M: in std_logic_vector(7 downto 0);
           C1, R1: in std_logic_vector(9 downto 0);
           rom_addr16: out std_logic_vector(15 downto 0);
           red : out std_logic_vector(2 downto 0);
           green : out std_logic_vector(2 downto 0);
           blue : out std_logic_vector(1 downto 0)
         );
end vga_ScreenSaver;

architecture vga_ScreenSaver of vga_ScreenSaver is
constant hbp: std_logic_vector(9 downto 0) := "0010010000";
    --Horizontal back porch = 144 (128+16)
constant vbp: std_logic_vector(9 downto 0) := "0000011111";
    --Vertical back porch = 31 (2+29)
constant w: integer := 240;
constant h: integer := 160;
signal xpix, ypix: std_logic_vector(9 downto 0);
signal spriteon: std_logic;

begin
    ypix <= vc - vbp - R1;
    xpix <= hc - hbp - C1;

    --Enable sprite video out when within the sprite region
    spriteon <= '1' when (((hc >= C1 + hbp) and (hc < C1 + hbp + w))
        and ((vc >= R1 + vbp) and (vc < R1 + vbp + h))) else '0';

    process(xpix, ypix)
        variable rom_addr1, rom_addr2: STD_LOGIC_VECTOR (16 downto 0);
    begin
        rom_addr1 := (ypix & "0000000") + ('0' & ypix & "000000")
            + ("00" & ypix & "00000") + ("000" & ypix & "0000");
                                        -- y*(128+64+32+16) = y*240
        rom_addr2 := rom_addr1 + ("00000000" & xpix);   -- y*240+x
        rom_addr16 <= rom_addr2(15 downto 0);
    end process;
```

Listing 6.10 (cont.) vga_ScreenSaver.vhd

```
    process(spriteon, vidon, M)
    variable j: integer;
    begin
            red <= "000";
            green <= "000";
            blue <= "00";
            if spriteon = '1' and vidon = '1' then
                    red <= M(7 downto 5);
                    green <= M(4 downto 2);
                    blue <= M(1 downto 0);
            end if;
    end process;

end vga_ScreenSaver;
```

We will implement the *bounce* algorithm using the method described in Chapter 3 (see Figs. 3.2 and 3.3). Listing 6.11 shows the VHDL code that implements the *bounce* component. Note that $C1$ and $R1$ will get updated on every rising edge of the clock *cclk*. If we use the 190 Hz clock *clk190* for this purpose the time for the photo to go from the top of the screen ($R1 = 0$) to the bottom of the screen $R1 = R1\max = 320$ will be $320/190 = 1.7s$, which will be reasonable.

In Listing 6.11 the first time that the input *go* goes high the variable *calc* is set to one. On the next rising edge of the clock the input *go* must have gone low so that from now on, with *calc* = 1, the statements corresponding to the *bounce* algorithm will be executed. This means that the *go* input must be a single pulse that lasts for one clock cycle when *btn*(0) is pressed. Our *clock_pulse* component from Example 3 will be used for this purpose as shown in Fig. 6.16. Note how the statements in Listing 6.11 implement the bounce algorithm given in Fig. 6.17.

Listing 6.12 is the VHDL program that implements the top-level design shown in Fig. 6.16.

Listing 6.11 vga_bounce.vhd

```vhdl
-- Example 38b: bounce
library IEEE;
use IEEE.STD_LOGIC_1164.all;
use IEEE.STD_LOGIC_unsigned.all;

entity bounce is
        port(
                cclk : in STD_LOGIC;
                clr : in STD_LOGIC;
                go : in STD_LOGIC;
                cl : out STD_LOGIC_VECTOR(9 downto 0);
                rl : out STD_LOGIC_VECTOR(9 downto 0)
             );
end bounce;

architecture bounce of bounce is
begin

process(cclk, clr)
variable clv, rlv: STD_LOGIC_VECTOR(9 downto 0);
variable dcv, drv: STD_LOGIC_VECTOR(9 downto 0);
variable calc: std_logic;
constant clmax: integer := 400;
constant rlmax: integer := 320;
begin
        if clr = '1' then
                clv := "0001010000";    -- 80 = 0x50
                rlv := "0010001100";    -- 140 = 0x8C
                dcv := "0000000001";    -- +1
                drv := "1111111111";    -- -1
                calc := '0';
        elsif cclk'event and cclk = '1' then
                if go = '1' then
                        calc := '1';
                elsif calc = '1' then
                        clv := clv + dcv;
                        rlv := rlv + drv;
                        if (clv < 0 or clv >= clmax) then
                                dcv := 0 - dcv;
                        end if;
                        if (rlv < 0 or rlv >= rlmax) then
                                drv := 0 - drv;
                        end if;
                end if;
        end if;
        cl <= clv;
        rl <= rlv;
end process;

end bounce;
```

Listing 6.12 vga_ScreenSaver_top.vhd

```vhdl
-- Example 38c: vga_ScreenSaver_top
library IEEE;
use IEEE.STD_LOGIC_1164.all;
use work.vga_components.all;

entity vga_ScreenSaver_top is
    port(
        mclk  : in STD_LOGIC;
        btn   : in STD_LOGIC_VECTOR(3 downto 0);
        hsync : out STD_LOGIC;
        vsync : out STD_LOGIC;
        red   : out std_logic_vector(2 downto 0);
        green : out std_logic_vector(2 downto 0);
        blue  : out std_logic_vector(1 downto 0)
        );
end vga_ScreenSaver_top;

architecture vga_ScreenSaver_top of vga_ScreenSaver_top is
signal clr, clk25, clk190, vidon, go1: std_logic;
signal hc, vc, C1, R1: std_logic_vector(9 downto 0);
signal M: std_logic_vector(7 downto 0);
signal rom_addr16: std_logic_vector(15 downto 0);
begin

    clr <= btn(3);

U1 : clkdiv2
    port map(mclk => mclk, clr => clr, clk190 => clk190,
        clk25 => clk25);

U2 : vga_640x480
    port map(clk => clk25, clr => clr, hsync => hsync,
        vsync => vsync, hc => hc, vc => vc, vidon => vidon);

U3 : vga_screensaver
    port map(vidon => vidon, hc => hc, vc => vc, M => M, C1 => C1,
        R1 => R1, rom_addr16 => rom_addr16, red => red,
        green => green, blue => blue);

U4 : loons240x160
        port map (addr => rom_addr16, clk => clk25, dout => M);

U5 : clock_pulse
    port map(inp => btn(0), cclk => clk190, clr => clr,
        outp => go1);

U6 : bounce
    port map(cclk => clk190, clr => clr, go => go1, c1 => C1,
        r1 => R1);

end vga_ScreenSaver_top;
```

Example 39

External Video RAM

In Example 31 we saw how to write data into and read data from the 16-Mbyte external RAM. In Example 32 we saw how to use the *MemUtil* application program from Digilent to store data in the 16-Mbyte external flash memory. You can use this same *MemUtil* application program to store data in the external RAM. In this example we will use *MemUtil* to store the image of the three loons shown in Fig. 6.18 into the external RAM and then use this RAM as a video RAM that will continually display its contents on a VGA screen. The size of this image is 640 x 480 pixels and so will fill our entire VGA screen. The image is available on the website www.lbebooks.com as the JPEG image file *loons640x480.jpg*.

To download this loon image to the external RAM using *MemUtil* we must convert the JPEG image file to a binary file in which each byte contains the color of one pixel using the encoding scheme we used in Example 37 (see Eq. (6.3)). The Matlab program shown in Listing 6.13 will do this. The binary file *loons640x480.exm* that is created with this Matlab function *IMG2ExtMem.m* will contain 640 x 480 = 307,200 bytes. The image *img2* produced by this Matlab function will be the degraded 8-bit color image shown in Fig. 6.19.

Figure 6.18
Loons photo (640 x 480) by Edie Haskell

Figure 6.19
8-bit color image produced by Listing 6.13

To load these 307,200 bytes from the file *loons640x480.exm* into the external RAM you must first program the FPGA with the bit file *onboardmemcfgJtagClk.bit* that you get when you download *MemUtil* from Digilent. You would then run *MemUtil* and click on the *Load RAM* tab as shown in Fig. 6.20. Enter the *File Name*, type 0 for the *File Start Location*, type 307200 for the *Length*, and click *Load*. The external RAM will now contain the 8-bit loon image shown in Fig. 6.19, and will retain the image as long as power is applied to the board. When power is removed all data in the RAM are lost.

Listing 6.13 IMG2ExtMem.m

```matlab
function img2 = IMG2ExtMem(imgfile, outfile)
% Convert a .jpg image to 8-bit color
% and save it as a binary file (.exm)
% for loading to external memory (ram or flash)
% each word contains two 8-bit pixels
% color byte: [R2,R1,R0,G2,G1,G0,B1,B0]
% img2 = Img2ExtMem(imgfile, outfile)
% img2 is 256-bit color image
% imgfile = input .jpg file
% outfile = output .exm file
% Example:
% img2 = IMG2ExtMem('loons640x480.jpg', 'loons640x480.exm');
img = imread(imgfile);
height = size(img, 1)
width = size(img, 2)
s = fopen(outfile,'wb');   %opens the output file
cnt = 0;
img2 = img;

for r=1:height
    for c=1:width
        cnt = cnt + 1;
        R = img(r,c,1);
        G = img(r,c,2);
        B = img(r,c,3);
        Rb = dec2bin(R,8);
        Gb = dec2bin(G,8);
        Bb = dec2bin(B,8);
        img2(r,c,1) = bin2dec([Rb(1:3) '00000']);
        img2(r,c,2) = bin2dec([Gb(1:3) '00000']);
        img2(r,c,3) = bin2dec([Bb(1:2) '000000']);
        Outbyte = [ Rb(1:3) Gb(1:3) Bb(1:2) ];
        fwrite(s,uint8(bin2dec(Outbyte)),'uint8');
    end
end
fclose(s);
```

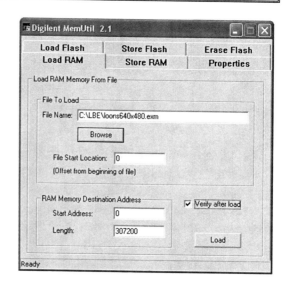

Figure 6.20 Downloading image data to external RAM using MemUtil

We will use the top-level design shown in Fig. 6.21 to display the image data in the external RAM on the VGA screen. Recall that we must display a VGA pixel at a 25 MHz rate, i.e., every 40 ns, but we can read a 16-bit word from memory only once every 80 ns. Therefore, each 16-bit word read from the RAM will contain two pixel data as shown in Fig. 6.22.

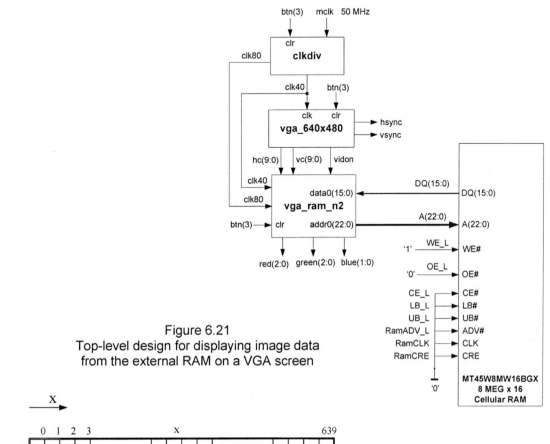

Figure 6.21
Top-level design for displaying image data from the external RAM on a VGA screen

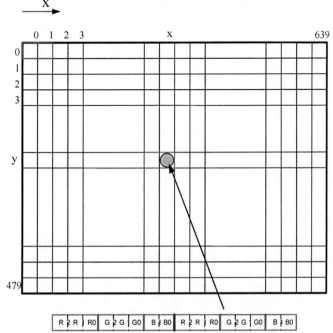

Figure 6.22 Each word in the video RAM contains color data for two pixels

178   Example 39

The video RAM shown in Fig. 6.22 will contain $640/2 = 320$ words/row or $320 \times 480 = 153,600$ words/screen. If a pixel is located at column x and row y in Fig. 6.22 then the word containing that pixel will be at address

$$addr0 = 320y + \frac{x}{2} = (256 + 64)y + \frac{x}{2} \qquad (6\text{-}4)$$

and the pixel value will be given by $pixel = x \bmod 2$, where $pixel$ = '0' is the most-significant byte and $pixel$ = '1' is the least-significant byte in the 16-bit word at address $addr0$.

The component *vga_ram_n2* shown in Fig. 6.21 needs to produce the address $addr0(22:0)$ from a counter that counts from 0 to 153,599 every 80 ns (*clk80*) and then displays each of the two pixels contained in $data0(15:0)$ for 40 ns (*clk40*) as shown in the timing diagram in Fig. 6.23. Listing 6.14 shows the VHDL program for the component *vga_ram_n2*.

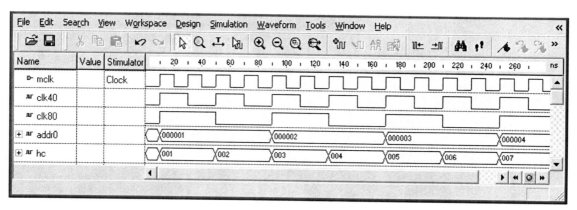

Figure 6.23 Timing associated with the component *vga_ram_n2* in Fig. 6.21

Listing 6.14 vga_ram_n2.vhd

```
-- Example 39a: vga_ram for Nexys-2 board
library IEEE;
use IEEE.STD_LOGIC_1164.ALL;
use IEEE.STD_LOGIC_UNSIGNED.ALL;

entity vga_ram_n2 is
    port ( clk40, clk80, clr : in std_logic;
           vidon: in std_logic;
           hc : in std_logic_vector(9 downto 0);
           vc : in std_logic_vector(9 downto 0);
           data0 : in std_logic_vector(15 downto 0);
           addr0 : out std_logic_vector(22 downto 0);
           red : out std_logic_vector(2 downto 0);
           green : out std_logic_vector(2 downto 0);
           blue : out std_logic_vector(1 downto 0)
    );
end vga_ram_n2;
```

Listing 6.14 (cont.) vga_ram_n2.vhd

```vhdl
architecture vga_ram_n2 of vga_ram_n2 is

constant addr_max: std_logic_vector(19 downto 0) := X"25800";
    --Max address = 320x480 = 153,600 = X"25800"
signal pixel: std_logic_vector(7 downto 0);
signal addr_count: std_logic_vector(19 downto 0);
begin
    --Counter for address bus - increment every 80 ns
    process(clk80, clr)
    begin
        if clr = '1' then
            addr_count <= (others => '0');
        elsif(clk80'event and clk80 = '1') then
            if addr_count = addr_max - 1 then
                --Reset when max address is reached
                addr_count <= (others => '0');
            else
                if vidon = '1' then
                    addr_count <= addr_count + 1; --Increment address
                end if;
            end if;
        end if;
    end process;

    --Get 2 pixels per address
    process(clk40, clr)
    begin
        if clr = '1' then
            pixel <= X"00";
        elsif(clk40'event and clk40 = '1') then
            if clk80 = '1' then
                --get high byte
                pixel <= data0(15 downto 8);
            else
                --get low byte
                pixel <= data0(7 downto 0);
            end if;
        end if;
    end process;

    process(vidon, pixel)
    begin
        red <= "000";
        green <= "000";
        blue <= "00";
        if vidon = '1' then
            red <= pixel(7 downto 5);
            green <= pixel(4 downto 2);
            blue <= pixel(1 downto 0);
        end if;
    end process;

    addr0 <= "000" & addr_count;
end vga_ram_n2;
```

Listing 6.15 gives the VHDL program for the top-level design shown in Fig. 6.21. Note in this example that we are only reading from the external RAM and thus the *WE_L* output is set to 1.

Listing 6.15 vga_extRam_top.vhd

```vhdl
-- Example 39b: vga_extRam_top
library IEEE;
use IEEE.STD_LOGIC_1164.all;
use work.vga_components.all;

entity vga_extRam_top is
      port(
            mclk : in STD_LOGIC;
            btn : in STD_LOGIC_VECTOR(3 downto 0);
            sw : in STD_LOGIC_VECTOR(7 downto 0);
            hsync : out STD_LOGIC;
            vsync : out STD_LOGIC;
            red : out std_logic_vector(2 downto 0);
            green : out std_logic_vector(2 downto 0);
            blue : out std_logic_vector(1 downto 0);
            A : out STD_LOGIC_VECTOR(22 downto 0);
            DQ : in STD_LOGIC_VECTOR(15 downto 0);
            CE_L : out STD_LOGIC;
            UB_L : out STD_LOGIC;
            LB_L : out STD_LOGIC;
            WE_L : out STD_LOGIC;
            OE_L : out STD_LOGIC;
            FlashCE_L : out STD_LOGIC;
            RamCLK : out STD_LOGIC;
            RamADV_L : out STD_LOGIC;
            RamCRE : out STD_LOGIC
         );
end vga_extRam_top;

architecture vga_extRam_top of vga_extRam_top is
signal clr, clk40, clk80, vidon: std_logic;
signal hc, vc: std_logic_vector(9 downto 0);
signal data0: std_logic_vector(15 downto 0);
signal addr0: std_logic_vector(22 downto 0);
begin

      clr <= btn(3);
      A <= addr0;
      FlashCE_L <= '1';         -- Disable Flash
      CE_L <= '0';              -- Enable ram
      UB_L <= '0';
      LB_L <= '0';
      RamCLK <= '0';
      RamADV_L <= '0';
      RamCRE <= '0';
      WE_L <= '1';              -- read only
      OE_L <= '0';              -- enable data bus
      data0 <= DQ;
```

Listing 6.15 (cont.) vga_extRam_top.vhd

```vhdl
U1 : clkdiv3
     port map(mclk => mclk, clr => clr, clk80 => clk80,
              clk40 => clk40);

U2 : vga_640x480
     port map(clk => clk40, clr => clr, hsync => hsync,
              vsync => vsync, hc => hc, vc => vc,
              vidon => vidon);

U3 : vga_ram_n2
     port map(clk40 => clk40, clk80 => clk80, clr => clr,
              vidon => vidon, hc => hc, vc => vc,
              data0 => data0, addr0 => addr0, red => red,
              green => green, blue => blue);

end vga_extRam_top;
```

Example 40

External Video Flash

The only problem with storing the loon photo in Example 39 in the external RAM is that when you turn off the power the RAM data vanishes. In this example we will store the same loon photo in the external flash memory and use that memory to continuously refresh the VGA screen. Our top-level design is shown in Fig. 6.24, which is very similar to Fig. 6.21. You will load the same binary file *loons640x480.exm* that you generated in Example 39 into the flash memory using the *MemUtil* application program as described in Examples 32 and 39.

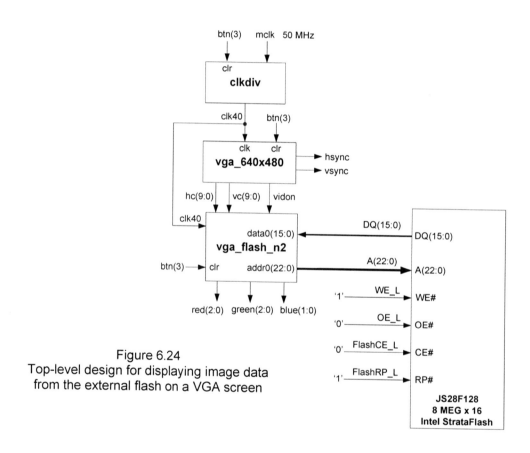

Figure 6.24
Top-level design for displaying image data from the external flash on a VGA screen

We saw in Example 32 that it will take up to 120 ns to read the first word from the flash memory. However, if this address is 0, it will buffer the four words at addresses 0, 1, 2, and 3. You can then access each of these next three words by waiting only 25 ns between reads rather than 120 ns. If you then access a word at another address on a 4-word boundary, you must wait 120 ns, which will then buffer the next four words. In Fig. 6.24 we read 16 bits (two pixels) at a time and we must send the pixels to the VGA screen at a 25 MHz rate, i.e., every 40 ns. We can meet both the flash memory timing and the VGA timing by reading four words (8 pixels) in $8 \times 40\text{ns} = 320\text{ns}$. The state diagram shown in Fig. 6.25 shows how we wait for four states ($4 \times 40\text{ns} = 160\text{ns}$) to read the first 16-bit value and store it in *px*1 shown in Fig. 6.26. We can then read the next three words every 40 ns and store the results in *px*2, *px*3, and *px*4 in Fig. 6.26. We then need to wait another 160 ns to read the next word.

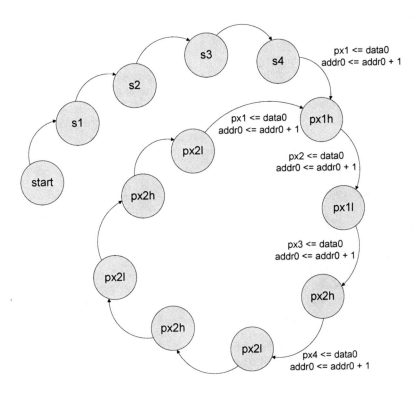

Figure 6.25 State diagram used to read data from the flash memory

px1	px1h	px1l
px2	px2h	px2l
px3	px3h	px3l
px4	px4h	px4l

Figure 6.26 Registers used to store the flash data using the state diagram in Fig. 6.25

The state diagram in Fig. 6.25 is implemented in the VHDL program for the component *vga_flash_n2* given in Listing 6.16. The top-level design shown in Fig. 6.24 is implemented by the VHDL program shown in Listing 6.17. Note that because the flash memory is an Intel chip the words are stored in memory in *little endian* fashion, i.e., the least-significant byte is stored in lower memory. This is opposite to the way the same image data were stored in the external RAM. This means that in Listing 6.17 we have to flip the bytes when assigning the input *DQ*(15:0) to the signal *data0*(15:0).

Listing 6.16 vga_flash_n2.vhd

```vhdl
-- Example 40a: vga_flash for Nexys-2 board
library IEEE;
use IEEE.STD_LOGIC_1164.ALL;
use IEEE.STD_LOGIC_UNSIGNED.ALL;

entity vga_flash_n2 is
    port ( clk40, clr : in std_logic;
            vidon: in std_logic;
            hc : in std_logic_vector(9 downto 0);
            vc : in std_logic_vector(9 downto 0);
            data0 : in std_logic_vector(15 downto 0);
            addr0 : out std_logic_vector(22 downto 0);
            red : out std_logic_vector(2 downto 0);
            green : out std_logic_vector(2 downto 0);
            blue : out std_logic_vector(1 downto 0)
    );
end vga_flash_n2;

architecture vga_flash_n2 of vga_flash_n2 is

signal pixel: std_logic_vector(7 downto 0);
signal px1,px2,px3, px4: std_logic_vector(15 downto 0);
signal addr_count: std_logic_vector(19 downto 0);
type state_type is (start, s1, s2, s3, s4, px1h, px1l,
            px2h, px2l,px3h, px3l,px4h, px4l);
signal state: state_type;
constant addr_max: std_logic_vector(19 downto 0) := X"25800";
   --Max address = 320x480 = 153,600 = X"25800"

begin

faddr: process(clk40, clr)
  begin
    if clr = '1' then
        state <= start;
        addr_count <= (others => '0');
        px1 <= (others => '0');
        px2 <= (others => '0');
        px3 <= (others => '0');
        px4 <= (others => '0');
    elsif (clk40'event and clk40 = '1') then
      if vidon = '1' then
        case state is
            when start =>
                state <= s1;
            when s1 =>
                state <= s2;
            when s2 =>
                state <= s3;
            when s3 =>
                state <= s4;
```

Listing 6.16 (cont.) vga_flash_n2.vhd

```vhdl
            when s4 =>
                state <= px1h;
                if addr_count = addr_max - 1 then
                    --Reset when max address is reached
                    addr_count <= (others => '0');
                else
                    if vidon = '1' then
                        addr_count <= addr_count + 1;
                        px1 <= data0;
                    end if;
                end if;
            when px1h =>
                state <= px1l;
                if addr_count = addr_max - 1 then
                    --Reset when max address is reached
                    addr_count <= (others => '0');
                else
                    if vidon = '1' then
                        addr_count <= addr_count + 1;
                    px2 <= data0;
                    end if;
                end if;
            when px1l =>
                state <= px2h;
                if addr_count = addr_max - 1 then
                    --Reset when max address is reached
                    addr_count <= (others => '0');
                else
                    if vidon = '1' then
                        addr_count <= addr_count + 1;
                    px3 <= data0;
                    end if;
                end if;
            when px2h =>
                state <= px2l;
                if addr_count = addr_max - 1 then
                    --Reset when max address is reached
                    addr_count <= (others => '0');
                else
                    if vidon = '1' then
                        addr_count <= addr_count + 1;
                    px4 <= data0;
                    end if;
                end if;
            when px2l =>
                state <= px3h;
            when px3h =>
                state <= px3l;
            when px3l =>
                state <= px4h;
            when px4h =>
                state <= px4l;
```

Listing 6.16 (cont.) vga_flash_n2.vhd

```vhdl
                when px4l =>
                    state <= px1h;
                    if addr_count = addr_max - 1 then
                        --Reset when max address is reached
                        addr_count <= (others => '0');
                    else
                        if vidon = '1' then
                            addr_count <= addr_count + 1;
                        px1 <= data0;
                        end if;
                    end if;
            end case;
        end if;
    end if;
  end process faddr;

C2: process(state,px1, px2, px3, px4)
begin
    pixel <= (others => '0');
    case state is
        when px1h =>
            pixel <= px1(15 downto 8);
        when px1l =>
            pixel <= px1(7 downto 0);
        when px2h =>
            pixel <= px2(15 downto 8);
        when px2l =>
            pixel <= px2(7 downto 0);
        when px3h =>
            pixel <= px3(15 downto 8);
        when px3l =>
            pixel <= px3(7 downto 0);
        when px4h =>
            pixel <= px4(15 downto 8);
        when px4l =>
            pixel <= px4(7 downto 0);
        when others =>
            null;
    end case;
end process;

process(vidon, pixel)
    begin
        red <= "000";
        green <= "000";
        blue <= "00";
        if vidon = '1' then
            red <= pixel(7 downto 5);
            green <= pixel(4 downto 2);
            blue <= pixel(1 downto 0);
        end if;
end process;

addr0 <= "000" & addr_count;
end vga_flash_n2;
```

Listing 6.17 vga_flash_n2_top.vhd

```vhdl
library IEEE;
use IEEE.STD_LOGIC_1164.all;
use IEEE.std_logic_unsigned.all;
use work.vga_components.all;

entity vga_flash_n2_top is
    port(
        mclk : in STD_LOGIC;
        btn : in STD_LOGIC_VECTOR(3 downto 0);
        hsync : out STD_LOGIC;
        vsync : out STD_LOGIC;
        red : out std_logic_vector(2 downto 0);
        green : out std_logic_vector(2 downto 0);
        blue : out std_logic_vector(1 downto 0);
        A : out STD_LOGIC_VECTOR(22 downto 0);
        DQ : in STD_LOGIC_VECTOR(15 downto 0);
        FlashCE_L : out STD_LOGIC;
        FlashRp_L: out STD_LOGIC;
        CE_L : out STD_LOGIC;
        WE_L : out STD_LOGIC;
        OE_L : out STD_LOGIC
    );
end vga_flash_n2_top;

architecture vga_flash_n2_top of vga_flash_n2_top is

    signal clr, clk40, vidon: std_logic;
    signal data0: std_logic_vector(15 downto 0);
    signal hc, vc: std_logic_vector(9 downto 0);
begin
    clr <= btn(3);
    -- Constant outputs
    FlashCE_L <= '0';           -- Enable flash
    flashRp_L <= '1';
    CE_L <= '1';                -- Disable RAM
    OE_L <= '0';
    WE_L <= '1';
    -- flash stores data little endian
    data0(15 downto 8) <= DQ(7 downto 0);
    data0(7 downto 0) <= DQ(15 downto 8);

U1 : clkdiv
    port map(mclk => mclk, clr => clr, clk40 => clk40);

U2 : vga_flash_n2
    port map(clk40 => clk40, clr => clr, vidon => vidon,
        hc => hc, vc => vc, data0 => data0,
        addr0 => A, red => red, green => green,
        blue => blue);

U3 : vga_640x480
    port map(clk => clk40, clr => clr, hsync => hsync,
        vsync => vsync, hc => hc, vc => vc,
        vidon => vidon);
end vga_flash_n2_top;
```

7

PS/2 Port

The PS/2 port is a 6-pin connector located at the left end of the slide switches on the Nexys-2 board as shown in Fig. 7.1. In addition to a power pin and a ground pin there is a bi-directional *clock* pin and a bi-directional *data* pin. The clock and data pins are connected to the FPGA pins *PS2C* and *PS2D* defined in the *nexys2.ucf* file. The PS/2 port is typically connected to some input device such as a keyboard or mouse. We will show how to interface this port to a keyboard in Example 41 and to a mouse in Example 42.

Figure 7.1 The PS2 port on the Nexys-2 board

Most of the time data are being transferred from the device to the host – the FPGA in this case. This is called *device-to-host communication*. If the host needs to send a command to the device it uses *host-to-device communication*. The device always generates the clock signal regardless of the direction of data flow. The clock frequency must be in the range 10 – 16.7 kHz. Data are sent one byte at a time in the form of an 11-bit frame containing a start bit, 8 data bits, a parity bit, and a stop bit.

The timing for device-to-host communication is shown in Fig. 7.2. In this case the device generates both the clock and the data signals. In the idle state both clock and data signals are high. The host will latch the data on the *falling* edge of the clock signal.

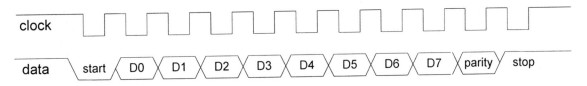

Figure 7.2 PS2 timing for device-to-host communication

Host-to-device communication is somewhat more complicated because the host has to ask the device to generate the clock signal. In addition, because the clock and data lines are bi-directional, you will need to include a tri-state buffer in each line as shown in Fig. 7.3. When the enable signal of a tri-state buffer is high the output is equal to the input; however, when the enable signal is low the output is high impedance. This is the way that the host can release the clock and/or data line so that the device can send data to the host. The VHDL code for the tri-state buffer is shown in Listing 7.1.

Listing 7.2 buff3.vhd
```vhdl
-- Example 41a: Tri-state buffer
library IEEE;
use IEEE.STD_LOGIC_1164.all;

entity buff3 is
    generic (N:integer);
    port(
        input : in STD_LOGIC_vector(N-1 downto 0);
        en : in STD_LOGIC;
        output : out STD_LOGIC_vector(N-1 downto 0)
    );
end buff3;


architecture buff3 of buff3 is
begin
        output <= input when en = '1' else (others => 'Z');
end buff3;
```

The clock and data signals sent from the device may be somewhat noisy signals. To read these signals accurately it is necessary to filter the signals as shown in Fig. 7.3. There are different ways you can do this. In Examples 41 and 42 we will shift the signals through a shift register at 25 MHz and require eight ones in a row for the signal to go high and eight zeros in a row for the signal to go low.

The timing for host-to-device communication is shown in Fig. 7.4. The host begins by bringing the clock signal low at time t_1 for at least 100 microseconds. At time t_2 the host brings the data line low. This is the *start* bit. The host then releases the clock line at time t_3 by bringing the enable line *cen* of the tri-state buffer in Fig. 7.3 low. This will cause the device to read the start bit. The host then waits for the device to bring the clock line low at time t_4. The host will then put the value of $D0$ on the data line *PS2D*. The device will latch this data on the *rising* edge of the clock and then bring the clock back low. On this falling edge of the clock the host will put the next value $D1$ on the data line and the device will latch it on the next rising edge of the clock. This process continues until the host has sent out eight data bits plus the parity bit (odd parity) plus the stop bit (1).

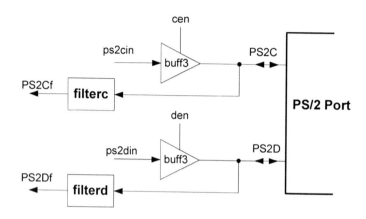

Figure 7.3 Interfacing to the PS/2 port

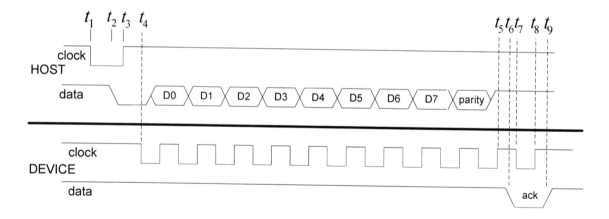

Figure 7.4 PS2 timing for host-to-device communication

After sending out the stop bit, which the device reads at time t_5, the host releases the data line by bringing the enable line *den* of the tri-state buffer in Fig. 7.3 low. The host then waits for the device to bring the data line low at time t_6 and the clock line low at time t_7. Finally, the host waits for the device to release the clock and data lines at times t_8 and t_9 respectively.

Example 41

Keyboard

In this example we will interface the PS/2 port to a PS/2 keyboard, also known as an AT keyboard. The example will not apply to the newer USB keyboards, or to the older, obsolete XT keyboard. Keyboards contain their own microprocessors that continually scan the keys and then send the resulting key pressings to the host – in our case through the PS/2 port.

For a PS/2 keyboard the key pressed is identified by a scan code. This code is associated with a physical key. Thus, the left shift key and the right shift key have different scan codes. When you press a key the *Make* scan code is sent to the PS/2 port. When you release the key the *Break* scan code is sent to the PS/2 port. The *Make* and *Break* scan codes for all the keys on a PS/2 keyboard are shown in Table 7.1.

For all of the letters and digits the *Make* scan code is a single byte and the *Break* scan code is the same byte preceded by the hex byte F0. Some of the other keys have a 2-byte *Make* scan code in which the first byte is hex E0. The corresponding *Break* scan code has three bytes starting with hex E0. Note that the *PrntScrn* and *Pause* keys are special with a 4-byte and 8-byte *Make* code respectively.

The scan codes have no relationship to the ASCII codes of the key characters. Recall that the ASCII code for an upper-case letter is different from the ASCII code of a lower-case letter. To tell if you are typing an upper-case or lower-case letter you would need to check if you are pressing the shift key (or if you have pressed the CapsLock key) and then press the letter key before you release the shift key. For example, if you want to type an upper-case A you would press the left shift key, press the key A, release the key A, and release the left shift key. From Table 7.1 this would send the following bytes to the PS/2 port.

```
12 1C F0 1C F0 12
```

When you hold a key down the *typematic* feature of the keyboard will, after a *typematic delay* of 0.25 – 1.00 seconds, continue to send out the *Make* scan code at a *typematic rate* of 2 – 30 characters per second. The typematic delay and rate can be changed by sending the 0xF3 command to the keyboard followed by a byte than encodes the new delay and rate.

In this example we will only read data from the keyboard and not send the keyboard any commands. Therefore, we don't need the tri-state buffers in Fig. 7.3. We will, however, need to filter the clock and data signals coming from the keyboard. The filtered data signal, *PS2Df*, will be shifted into two 11-bit words as shown in Fig. 7.5. Note that after shifting in these two words the first byte shifted in will be in *shift*2(8:1) and the second byte shifted in will be in *shift*1(8:1).

Listing 7.2 shows the VHDL code for interfacing to the keyboard. The output *xkey*(15:0) will contain the two bytes shifted in when a key is pressed on the keyboard.

Listing 7.3 is the VHDL program for the top-level design that will display the scan codes of the keys pressed on the 7-segment display.

Table 7.1 Keyboard Scan Codes

Key	Make	Break	Key	Make	Break	Key	Make	Break
A	1C	F0,1C	`	0E	F0,0E	F1	05	F0,05
B	32	F0,32	-	4E	F0,4E	F2	06	F0,06
C	21	F0,21	=	55	F0,55	F3	04	F0,04
D	23	F0,23	\	5D	F0,5D	F4	0C	F0,0C
E	24	F0,24	BKSP	66	F0,66	F5	03	F0,03
F	2B	F0,2B	SPACE	29	F0,29	F6	0B	F0,0B
G	34	F0,34	TAB	0D	F0,0D	F7	83	F0,83
H	33	F0,33	CAPS	58	F0,58	F8	0A	F0,0A
I	43	F0,43	L Shift	12	F0,12	F9	01	F0,01
J	3B	F0,3B	R Shift	59	F0,59	F10	09	F0,09
K	42	F0,42	L Ctrl	14	F0,14	F11	78	F0,78
L	4B	F0,4B	R Ctrl	E0,14	E0,F0,14	F12	07	F0,07
M	3A	F0,3A	L Alt	11	F0,11	Num	77	F0,77
N	31	F0,31	R Alt	E0,11	E0,F0,11	KP /	E0,4A	E0,F0,4A
O	44	F0,44	L GUI	E0,1F	E0,F0,1F	KP *	7C	F0,7C
P	4D	F0,4D	R GUI	E0,27	E0,F0,27	KP -	7B	F0,7B
Q	15	F0,15	Apps	E0,2F	E0,F0,2F	KP +	79	F0,79
R	2D	F0,2D	Enter	5A	F0,5A	KP EN	E0,5A	E0,F0,5A
S	1B	F0,1B	ESC	76	F0,76	KP .	71	F0,71
T	2C	F0,2C	Scroll	7E	F0,7E	KP 0	70	F0,70
U	3C	F0,3C	Insert	E0,70	E0,F0,70	KP 1	69	F0,69
V	2A	F0,2A	Home	E0,6C	E0,F0,6C	KP 2	72	F0,72
W	1D	F0,1D	Page Up	E0,7D	E0,F0,7D	KP 3	7A	F0,7A
X	22	F0,22	Page Dn	E0,7A	E0,F0,7A	KP 4	6B	F0,6B
Y	35	F0,35	Delete	E0,71	E0,F0,71	KP 5	73	F0,73
Z	1A	F0,1A	End	E0,69	E0,F0,69	KP 6	74	F0,74
0	45	F0,45	[	54	F0,54	KP 7	6C	F0,6C
1	16	F0,16	]	5B	F0,5B	KP 8	75	F0,75
2	1E	F0,1E	;	4C	F0,4C	KP 9	7D	F0,7D
3	26	F0,26	'	52	F0,52	U Arrow	E0,75	E0,F0,75
4	25	F0,25	,	41	F0,41	L Arrow	E0,6B	E0,F0,6B
5	2E	F0,2E	.	49	F0,49	D Arrow	E0,72	E0,F0,72
6	36	F0,36	/	4A	F0,4A	R Arrow	E0,74	E0,F0,74
7	3D	F0,3D	PrntScrn	E0,7C, E0,12	E0,F0,7C, E0,F0,12	Pause	E1,14, 77,E1, F0,14, F0,77,	None
8	3E	F0,3E						
9	46	F0,46						

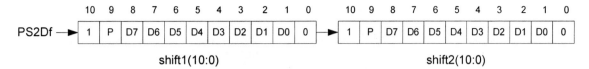

Figure 7.5 Shifting PS2Df into two bytes

Listing 7.2 keyboard.vhd

```vhdl
-- Example 41a: keyboard
library IEEE;
use IEEE.STD_LOGIC_1164.all;

entity keyboard is
    port(
          clk25 : in STD_LOGIC;
          clr : in STD_LOGIC;
          PS2C : in STD_LOGIC;
          PS2D : in STD_LOGIC;
          xkey : out STD_LOGIC_VECTOR(15 downto 0)
         );
end keyboard;


architecture keyboard of keyboard is
signal PS2Cf, PS2Df: std_logic;
signal ps2c_filter, ps2d_filter: std_logic_vector(7 downto 0);
signal shift1, shift2: std_logic_vector(10 downto 0);
begin

xkey <= shift2(8 downto 1) & shift1(8 downto 1);

-- filter for PS2 clock and data
filter: process(clk25, clr)
begin
   if clr = '1' then
        ps2c_filter <= (others => '0');
        ps2d_filter <= (others => '0');
        PS2Cf <= '1';
        PS2Df <= '1';
   elsif clk25'event and clk25 = '1' then
        ps2c_filter(7) <= PS2C;
        ps2c_filter(6 downto 0) <= ps2c_filter(7 downto 1);
        ps2d_filter(7) <= PS2D;
        ps2d_filter(6 downto 0) <= ps2d_filter(7 downto 1);
        if ps2c_filter = X"FF" then
             PS2Cf <= '1';
        elsif ps2c_filter = X"00" then
             PS2Cf <= '0';
        end if;
        if ps2d_filter = X"FF" then
             PS2Df <= '1';
        elsif ps2d_filter = X"00" then
             PS2Df <= '0';
        end if;
   end if;
end process filter;
```

Listing 7.2 (cont.) keyboard.vhd

```vhdl
--Shift Registers used to clock in scan codes from PS2--
shift: process(PS2Cf, clr)
begin
   if (clr = '1') then
        Shift1 <= (others => '0');
        Shift2 <= (others => '0');
   elsif (PS2Cf'event and PS2Cf = '0') then
        Shift1 <= PS2Df & Shift1(10 downto 1);
        Shift2 <= Shift1(0) & Shift2(10 downto 1);
   end if;
end process shift;

end keyboard;
```

Listing 7.3 keyboard_top.vhd

```vhdl
-- Example 41b: keyboard_top
library IEEE;
use IEEE.STD_LOGIC_1164.all;
use work.ps2_components.all;

entity keyboard_top is
    port(
        mclk : in STD_LOGIC;
        PS2C: in STD_LOGIC;
        PS2D: in STD_LOGIC;
        btn : in STD_LOGIC_VECTOR(3 downto 0);
        a_to_g : out STD_LOGIC_VECTOR(6 downto 0);
        dp : out STD_LOGIC;
        an : out STD_LOGIC_VECTOR(3 downto 0)
        );
end keyboard_top;

architecture keyboard_top of keyboard_top is
signal pclk, clk25, clk190, clr: std_logic;
signal xkey: std_logic_vector(15 downto 0);
begin
   clr <= btn(3);
   dp <= '1';         -- decimal points off

U1 : clkdiv2
   port map(mclk => mclk, clr => clr, clk25 => clk25,
        clk190 => clk190);

U2 : keyboard
   port map(clk25 => clk25, clr => clr, PS2C => PS2C,
        PS2D => PS2D, xkey => xkey);

U3 : x7segb
   port map(x => xkey, cclk => clk190, clr => clr,
        a_to_g => a_to_g, an => an);

end keyboard_top;
```

Example 42

Mouse

In this example we will create a mouse controller to interface a standard mouse to the PS/2 port. The mouse starts by entering a self-test mode when the power is turned on. After performing the self test, a 0xAA is sent to the host, followed by a device ID of 0x00. Communication between the mouse and the FPGA (host) uses the same 11-bit PS/2 data packet shown in Fig. 7.2. The mouse generates the clock at a frequency between 10 kHz and 16.7 kHz when data are transmitted between the mouse and the FPGA.

The mouse has four basic modes of operation: (1) *Reset mode*, the initial mode where the mouse performs initialization and self-tests, (2) *Stream mode* where the mouse transmits three data packets when movement occurs or the state of a button changes, (3) *Remote mode* where the host requests movement data packets, and (4) *Wrap mode* which is a diagnostic mode where the mouse echoes packets that it receives back to the host. By default, the mouse is in stream mode but does not transmit data. The host must transmit a 0xF4 to the mouse to initiate data reporting. The mouse will acknowledge by transmitting a 0xFA back to the host followed by movement data packets.

Movement data packets are three 11-bit data packets containing a mouse status byte followed by a byte containing *x*-movement data, and finally a byte containing *y*-movement data. These three 11-bit data packets can be shifted into the three shift registers *shift1*(10:0), *shift2*(10:0), and *shift3*(10:0) as shown in Figure 7.6.

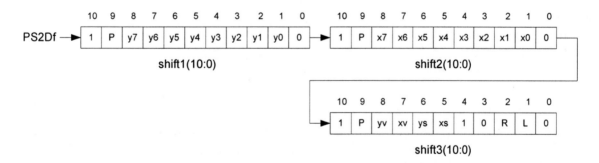

Figure 7.6 Shifting mouse data into three bytes

The *y*-movement byte is in *shift1*(8:1) and the *x*-movement byte is in *shift2*(8:1). This movement data represents the *velocity* with which the mouse is moved in the plus or minus *x* or *y* direction determined by the sign bits *xs* and *ys* located in the status byte at *shift3*(5) and *shift3*(6) respectively. Thus, the movement data are really 9-bit two's complement numbers where the most significant bit is the sign bit located in the mouse status byte. This gives the range of −256 to +255 for each of the movement directions. If this range is exceeded, the respective overflow bit *xv* or *yv* in the mouse status byte is set. The bit *L* in *shift*3(1) in Fig. 7.6 is set if the left mouse button is being pressed, and the bit *R* in *shift*3(2) is set if the right mouse button is being pressed. In streaming mode with data reporting turned on, the mouse continues to transmit these three packets, one after another, as the mouse moves or buttons are pressed.

There are many other commands in the command set for the mouse. For example, a 0xF5 will disable data reporting, a 0xF3 initiates a transmission sequence for setting the mouse data sampling rate, and a 0xE8 will start a transmission sequence for setting the resolution, or number of millimeters per count that the mouse uses to generate the *x*- and *y*-movement data. All commands sent to the mouse are met with a 0xFA acknowledge byte transmitted back to the host. The mouse may also send a 0xFE to request a data resend from the host, or a 0xFC signifying an error.

Fig. 7.7 shows the state diagram for a mouse controller and Listing 7.4 shows the VHDL code that implements this state diagram.

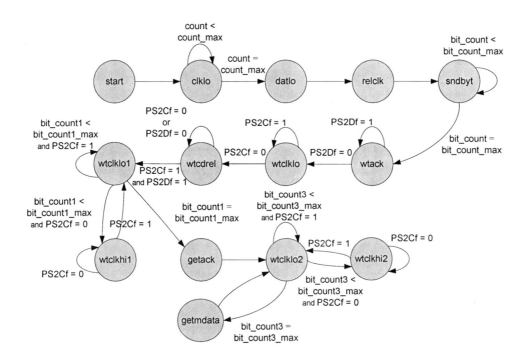

Figure 7.7 State diagram for a mouse controller

Execution begins in the start state where the clock enable signal *cen* is brought high to give the host control of the clock line. The signal *ps2cin* is reset to zero pulling the clock line low. This is the first step for initiating host-to-device communication. The next state counts for 2,500 clock cycles waiting 100 microseconds, the time required for the host to hold the clock low. After holding for 100 microseconds, *den* is set to enable data output, the clock is released and a byte of data, 0xF4, followed by the parity and stop bits are sent in the *sndbyt* state. Following the transmission, the *wtack* state waits for the data line to go low followed by the clock going low in *wtclklo* for the active low acknowledge pulse given by the mouse. After this acknowledge pulse has been received, the state machine shifts the acknowledge byte 0xFA from the mouse into the host. Because the mouse needs an extra clock cycle when it switches from reading the stop bit from the host to setting the start bit for transmitting the 0xFA acknowledge byte, the *wtclklo1 – wtclkhi1* states will need to clock in 12 bits to get the 0xFA acknowledge byte, which it loads into *y_mouse* for display purposes. After the clock goes high, the *wtclklo2* and *wtclkhi2* states wait for the clock to pulse low and then high again starting the transmission of the 3 mouse data bytes. After the clock goes low, data are shifted into the host in *wtclklo2*. Execution continues looping between *wtclklo2* and *wtclkhi2* shifting in 33 times since each of the 3 bytes is an 11-bit packet. After shifting data in 33 times, the 9-bit velocity values are stored in *x_mouse_v*(8:0) and *y_mouse_v*(8:0), and the status byte is stored in *byte3*(7:0). In the *getmdata* state 9-bit distance values *x_mouse_d*(8:0) and *y_mouse_d*(8:0) are calculated by keeping a running sum of the velocity values. This process of collecting the mouse movement data continues between *wtclklo2*, *wtclkhi2*, and *getmdata* indefinitely.

The outputs of the *mouse_ctrl* component *x_data*(9:0) and *y_data*(9:0) will be either the velocity data or the distance data depending on the value of the input *sel* using the multiplexer described by the *outsel* process at the end of Listing 7.4. The top-level design shown in Listing 7.5 can be used to test the mouse controller by displaying the velocity data on the 7-segment display or the distance data if *btn*(0) is pressed. The LEDs will display the bits in *byte3*.

Listing 7.4 mouse_ctrl.vhd

```vhdl
-- Example 42a: mouse_ctrl
library IEEE;
use IEEE.STD_LOGIC_1164.all;
use IEEE.STD_LOGIC_unsigned.all;

entity mouse_ctrl is
     port(
            clk25 : in STD_LOGIC;
            clr : in STD_LOGIC;
            sel : in STD_LOGIC;
            PS2C : inout STD_LOGIC;
            PS2D : inout STD_LOGIC;
            byte3 : out STD_LOGIC_VECTOR(7 downto 0);
            x_data : out STD_LOGIC_VECTOR(8 downto 0);
            y_data : out STD_LOGIC_VECTOR(8 downto 0)
          );
end mouse_ctrl;

architecture mouse_ctrl of mouse_ctrl is
type state_type is (start, clklo, datlo, relclk, sndbyt, wtack,
        wtclklo, wtcdrel, wtclklo1, wtclkhi1, getack, wtclklo2,
        wtclkhi2, getmdata);
signal state: state_type;
signal PS2Cf, PS2Df, cen, den, sndflg, xs, ys: std_logic;
signal ps2cin, ps2din, ps2cio, ps2dio: std_logic;
signal ps2c_filter, ps2d_filter: std_logic_vector(7 downto 0);
```

Listing 7.4 (cont.) mouse_ctrl.vhd

```vhdl
    signal x_mouse_v, y_mouse_v: std_logic_vector(8 downto 0);
    signal x_mouse_d, y_mouse_d: std_logic_vector(8 downto 0);
    signal shift1, shift2, shift3: std_logic_vector(10 downto 0);
    signal f4cmd: std_logic_vector(9 downto 0);
    signal bit_count: std_logic_vector(3 downto 0);
    signal bit_count1: std_logic_vector(3 downto 0);
    signal bit_count3: std_logic_vector(5 downto 0);
    signal count: std_logic_vector(11 downto 0);
    constant count_max: std_logic_vector(11 downto 0) := X"9C4";
                                                       -- 2500 100 us
    constant bit_count_max: std_logic_vector(3 downto 0) := "1010";
                                                       -- 10
    constant bit_count1_max: std_logic_vector(3 downto 0) := "1100";
                                                       -- 12 ack
    constant bit_count3_max: std_logic_vector(5 downto 0) := "100001";
                                                       -- 33
begin

-- tri-state buffers
ps2cio <= ps2cin when cen = '1' else 'Z';
ps2dio <= ps2din when den = '1' else 'Z';
PS2C <= ps2cio;
PS2D <= ps2dio;

-- filter for PS2 clock
filter: process(clk25, clr)
begin
    if clr = '1' then
          ps2c_filter <= (others => '0');
          ps2d_filter <= (others => '0');
          PS2Cf <= '1';
          PS2Df <= '1';
    elsif clk25'event and clk25 = '1' then
          ps2c_filter(7) <= ps2cio;
          ps2c_filter(6 downto 0) <= ps2c_filter(7 downto 1);
          ps2d_filter(7) <= ps2dio;
          ps2d_filter(6 downto 0) <= ps2d_filter(7 downto 1);
          if ps2c_filter = X"FF" then
                PS2Cf <= '1';
          elsif ps2c_filter = X"00" then
                PS2Cf <= '0';
          end if;
          if ps2d_filter = X"FF" then
                PS2Df <= '1';
          elsif ps2d_filter = X"00" then
                PS2Df <= '0';
          end if;
    end if;
end process filter;
```

Listing 7.4 (cont.) mouse_ctrl.vhd

```vhdl
-- State machine for reading mouse
smouse: process(clk25, clr)
begin
   if (clr = '1') then
          state <= start;
          cen <= '0';
          den <= '0';
          ps2cin <= '0';
          count <= (others => '0');
          bit_count3 <= (others => '0');
          bit_count1 <= (others => '0');
          Shift1 <= (others => '0');
          Shift2 <= (others => '0');
          Shift3 <= (others => '0');
          x_mouse_v <= (others => '0');
          y_mouse_v <= (others => '0');
          x_mouse_d <= (others => '0');
          y_mouse_d <= (others => '0');
          sndflg <= '0';
   elsif (clk25'event and clk25 = '1') then
     case state is
        when start =>
              cen <= '1';           -- enable clock output
              ps2cin <= '0';  -- start bit
              count <= (others => '0'); -- reset count
              state <= clklo;
        when clklo =>
              if count < count_max then
                   count <= count + 1;
                   state <= clklo;
              else
                   state <= datlo;
                   den <= '1';             -- enable data output
              end if;
        when datlo =>
              state <= relclk;
              cen <= '0';         -- release clock
        when relclk =>
              sndflg <= '1';
              state <= sndbyt;
        when sndbyt =>
              if bit_count < bit_count_max then
                   state <= sndbyt;
              else
                   state <= wtack;
                   sndflg <= '0';
                   den <= '0';           -- release data
              end if;
        when wtack =>                              -- wait for data low
              if PS2Df = '1' then
                   state <= wtack;
              else
                   state <= wtclklo;
              end if;
```

Listing 7.4 (cont.) mouse_ctrl.vhd

```vhdl
            when wtclklo =>                       -- wait for clock low
                 if PS2Cf = '1' then
                      state <= wtclklo;
                 else
                      state <= wtcdrel;
                 end if;
            when wtcdrel =>            -- wait to release clock and data
                 if PS2Cf = '1' and PS2Df = '1' then
                      state <= wtclklo1;
                      bit_count1 <= (others => '0');
                 else
                      state <= wtcdrel;
                 end if;
            when wtclklo1 =>                      -- wait for clock low
                 if bit_count1 < bit_count1_max then
                    if PS2Cf = '1' then
                       state <= wtclklo1;
                    else
                       state <= wtclkhi1;         -- get ack byte FA
                       Shift1 <= PS2Df & Shift1(10 downto 1);
                    end if;
                 else
                    state <= getack;
                 end if;
            when wtclkhi1 =>                     -- wait for clock high
                 if PS2Cf = '0' then
                      state <= wtclkhi1;
                 else
                      state <= wtclklo1;
                      bit_count1 <= bit_count1 + 1;
                 end if;
            when getack =>                        -- get ack FA
                 y_mouse_v <= shift1(9 downto 1);
                 x_mouse_v <= shift2(8 downto 0);
                 byte3 <= shift1(10 downto 5) & shift1(1 downto 0);
                 state <= wtclklo2;
                 bit_count3 <= (others => '0');
            when wtclklo2 =>                      -- wait for clock low
              if bit_count3 < bit_count3_max then
                 if PS2Cf = '1' then
                    state <= wtclklo2;
                 else
                    state <= wtclkhi2;
                    Shift1 <= PS2Df & Shift1(10 downto 1);
                    Shift2 <= Shift1(0) & Shift2(10 downto 1);
                    Shift3 <= Shift2(0) & Shift3(10 downto 1);
                 end if;
              else
                 x_mouse_v <= shift3(5) & shift2(8 downto 1);  -- x vel
                 y_mouse_v <= shift3(6) & shift1(8 downto 1);  -- y vel
                 byte3 <= shift3(8 downto 1);
                 state <= getmdata;
              end if;
```

Listing 7.4 (cont.) mouse_ctrl.vhd

```vhdl
            when wtclkhi2 =>                    -- wait for clock high
                if PS2Cf = '0' then
                    state <= wtclkhi2;
                else
                    state <= wtclklo2;
                    bit_count3 <= bit_count3 + 1;
                end if;
            when getmdata =>     -- read mouse data and keep going
                x_mouse_d <= x_mouse_d + x_mouse_v;   -- x distance
                y_mouse_d <= y_mouse_d + y_mouse_v;   -- y distance
                bit_count3 <= (others => '0');
                state <= wtclklo2;
        end case;
    end if;
end process smouse;

-- Send F4 command to mouse
sndf4: process(PS2Cf, clr, sndflg)
begin
    if (clr = '1') then
        f4cmd <= "1011110100";   -- stop-parity-F4
        ps2din <= '0';
        bit_count <= (others => '0');
    elsif (PS2Cf'event and PS2Cf = '0' and sndflg = '1') then
        ps2din <= f4cmd(0);
        f4cmd(8 downto 0) <= f4cmd(9 downto 1);
        f4cmd(9) <= '0';
        bit_count <= bit_count + 1;
    end if;
end process sndf4;

-- Output select
outsel: process(x_mouse_v,y_mouse_v,x_mouse_d,y_mouse_d, sel)
begin
    if sel = '0' then
        x_data <= x_mouse_v;
        y_data <= y_mouse_v;
    else
        x_data <= x_mouse_d;
        y_data <= y_mouse_d;
    end if;
end process outsel;
end mouse_ctrl;
```

Listing 7.5 mouse_top.vhd

```vhdl
-- Example 42b: mouse_top
library IEEE;
use IEEE.STD_LOGIC_1164.all;
use work.ps2_components.all;

entity mouse_top is
      port(
            mclk : in STD_LOGIC;
            PS2C: inout STD_LOGIC;
            PS2D: inout STD_LOGIC;
            btn : in STD_LOGIC_VECTOR(3 downto 0);
            ld : out STD_LOGIC_VECTOR(3 downto 0);
            a_to_g : out STD_LOGIC_VECTOR(6 downto 0);
            dp : out STD_LOGIC;
            an : out STD_LOGIC_VECTOR(3 downto 0)
            );
end mouse_top;

architecture mouse_top of mouse_top is
signal clk25, clk190, clr: std_logic;
signal byte3: std_logic_vector(7 downto 0);
signal x_data, y_data: std_logic_vector(8 downto 0);
signal xmouse: std_logic_vector(15 downto 0);
begin
      clr <= btn(3);
      dp <= '1';           -- decimal points off
      xmouse <= x_data(7 downto 0) & y_data(7 downto 0);
      ld(0) <= y_data(8);
      ld(1) <= x_data(8);
      ld(2) <= byte3(1);              -- right button
      ld(3) <= byte3(0);              -- left button

U1 : clkdiv2
      port map(mclk => mclk, clr => clr, clk25 => clk25,
               clk190 => clk190);

U2 : mouse_ctrl
      port map(clk25 => clk25, clr => clr, sel => btn(0),
              PS2C => PS2C, PS2D => PS2D, byte3 => byte3,
              x_data => x_data, y_data => y_data);

U3 : x7segb
      port map(x => xmouse, cclk => clk190, clr => clr,
               a_to_g => a_to_g, an => an);

end mouse_top;
```

Listing 7.6 is a second top-level design that modifies the VGA sprite program from Example 36 by using the mouse to move the location of the initials on the screen.

Listing 7.6 vga_mouse_top.vhd

```vhdl
-- Example 42c: vga_mouse_top
library IEEE;
use IEEE.STD_LOGIC_1164.all;
use IEEE.STD_LOGIC_unsigned.all;
use work.ps2_components.all;

entity vga_mouse_top is
    port(
         mclk : in STD_LOGIC;
         btn : in STD_LOGIC_VECTOR(3 downto 0);
         PS2C : inout STD_LOGIC;
         PS2D : inout STD_LOGIC;
         ld : out std_logic_vector(1 downto 0);
         hsync : out STD_LOGIC;
         vsync : out STD_LOGIC;
         red : out std_logic_vector(2 downto 0);
         green : out std_logic_vector(2 downto 0);
         blue : out std_logic_vector(1 downto 0)
        );
end vga_mouse_top;

architecture vga_mouse_top of vga_mouse_top is
signal clr, clk25, clk190, vidon, sel: std_logic;
signal hc, vc: std_logic_vector(9 downto 0);
signal x_data, y_data: std_logic_vector(8 downto 0);
signal M: std_logic_vector(0 to 31);
signal rom_addr4: std_logic_vector(3 downto 0);
signal byte3 : STD_LOGIC_VECTOR(7 downto 0);
signal cursor_row, cursor_col: STD_LOGIC_VECTOR(9 downto 0);
constant xc: STD_LOGIC_VECTOR(9 downto 0) := "0101000000"; -- 320
constant yc: STD_LOGIC_VECTOR(9 downto 0) := "0011110000"; -- 240

begin
   clr <= btn(3);
   ld(0) <= byte3(1);              -- right button
   ld(1) <= byte3(0);              -- left button
   sel <= '1';                     -- distance data
   cursor_row <= yc - (y_data(8) & y_data);
   cursor_col <= xc + (x_data(8) & x_data);

U1 : clkdiv2
   port map(mclk => mclk, clr => clr, clk25 => clk25,
         clk190 => clk190);

U2 : vga_640x480
   port map(clk => clk25, clr => clr, hsync => hsync,
         vsync => vsync, hc => hc, vc => vc, vidon => vidon);

U3 : vga_mouse_initials
   port map(vidon => vidon, hc => hc, vc => vc, M => M,
         cursor_row => cursor_row, cursor_col => cursor_col,
         rom_addr4 => rom_addr4, red => red, green => green,
         blue => blue);
```

Listing 7.6 (cont.) vga_mouse_top.vhd
```
U4 : prom_dmh
   port map(addr => rom_addr4, M => M);

U5 : mouse_ctrl
   port map(clk25 => clk25, clr => clr, sel => sel, PS2C => PS2C,
        PS2D => PS2D, byte3 => byte3, x_data => x_data,
        y_data => y_data);

end vga_mouse_top;
```

Test Benches

So far, we have tested our designs by simulating the hardware and inspecting the output generated given the input stimulators. After using the simulator to debug the hardware and verifying that the design works as expected for the test conditions applied, we have typically downloaded the design to the Spartan3E FPGA on the Nexys-2 board for final testing.

For designs where there are numerous inputs or the input data have significant timing considerations, writing a test bench to ensure that the hardware designed behaves correctly can be easier than setting stimulators. Test benches are also useful for testing many different input test cases. It is particularly useful when you need to simulate the behavior of some external device such as a mouse, keyboard, or external RAM.

In this example, we designed a mouse controller that communicates with the mouse by sending a 0xF4 to enable the data reporting mode, accepts the mouse's 0xFA acknowledging the request, and processes the continuous stream of three 11-bit mouse movement data packets. This communication is based upon a clock generated by the mouse at a frequency of 10-16.7 kHz, much lower than the boards 50 MHz clock. Communication takes place in two modes, device-to-host and host-to-device, with each having different timing and signal sequencing specifications. Given these complexities, simulating the mouse controller would require great planning in setting up stimulators on the simulation waveform and therefore, a test bench is more realistic and effective.

The full version of Aldec's Active-HDL provides a tool that generates a basic test bench file for any component instead of typing one from scratch. Appendix B shows the steps to do this; however, if you have the Student Edition, you will have to type the test bench starting from a blank *.vhd* file. First, we must identify which component is going to be the *unit under test* (UUT). In larger designs, hardware designers may write test benches for individual components. We will test our mouse controller by writing a test bench that behaves the same as the mouse and will interact with the top-level design. There is no need to test individual components since there are so few components used to implement the mouse controller. Therefore, our UUT will be the *mouse_top* entity shown in Listing 7.5.

Listing 7.7 shows the test bench, *mouse_top_tb*. The standard library and our *ps2_components* library are included. The *mouse_top_tb* entity is defined with no inputs or outputs, instead, *mouse_top_tb* will generate the inputs to the UUT and the outputs will be observed on a waveform. The architecture contains the component declaration for the UUT, in this case *mouse_top*. Signals are declared for each of the UUT's inputs and outputs for port mapping the UUT into the test bench. To create a test bench, we will write VHDL code from the mouse's perspective. A signal, *m_data_in,* is declared for storing data received from the host. Finally, a signal *END_SIM* is used for setting the end of the simulation to stop ongoing signals, such as the clock.

Processes for stimulus signals follow the port map for the UUT. The first process generates a 50 MHz clock signal by inverting *mclk* every 10 ns until the *END_SIM* signal goes high. *btn(3)* is set high to clear the mouse controller. *btn(0)* is reset to zero to output velocity data. Upon power up, the mouse performs a series of self tests and sends a 0xAA to the host followed by a device ID of 0x00, assuming the mouse is working properly. Using the *wait* statement

```
wait for 1 us;
```

the test bench simulates a 1 microsecond self test operation. After the self test is complete, the mouse sends 0xAA by generating a 16 kHz clock and transmitting the 11 bit packet containing a start bit, data byte least significant-bit first, parity bit, and stop bit. A 16 kHz clock has a period of 62.5 microseconds and therefore switches from high to low every 31.25 microseconds. As the clock pulses, data are shifted out the *PS2D* line. This process is repeated to send the device ID 0x00.

After transmitting the device ID, the mouse relinquishes the *PS2C* and *PS2D* lines to high impedance. At this point, we have enabled the mouse controller by resetting *btn(3)* to zero. Our test bench holds this clear signal high while the mouse transmits the 0xAA and 0x00 since our mouse controller is active well after the mouse has powered up and therefore does not receive the 0xAA and 0x00.

The mouse waits for the host to bring the clock and data lines low preparing for a host-to-device transmission:

```
wait until PS2C = '0';   --wait until clk line is held low by host
wait until PS2D = '0';   --wait until data line is low
```

The *wait until* statement waits until a signal meets the condition specified. After these signals have been asserted low by the host, the mouse must generate a 16 kHz clock and shift the data sent by the host on *PS2D* into an internal register in the mouse, *m_data_in*. After shifting in 11 bits, the mouse pulls the data line low signaling as an acknowledgement. Then, the data are validated by checking the parity bit and if the transmission is considered successful, the mouse transmits a 0xFA acknowledgement byte. Since our controller was designed under the assumption that the mouse and all communications are working properly, there is no need to write code within the test bench to transmit a mouse error code. Instead, we can use the *assert* command to validate that the communication between our UUT and mouse is commencing as expected. The general form of the *assert* command is shown below.

```
assert <condition>
report "<report_string>"
[severity {note|warning|error|failure}];
```

where *severity* is optional. A severity of *failure* will stop the simulation. The parity bit is checked and using the *assert* statement, simulation will stop and a failure will be reported if the parity is not odd.

```
cnt := 0;
for x in 9 downto 1 loop
     if m_data_in(x) = '1' then
          cnt := cnt + 1;
     end if;
end loop;

assert (cnt = 1 or cnt = 3 or cnt = 5 or cnt = 7 or cnt = 9)
report "Parity bit incorrect"
severity failure;

report "Data received";
```

The *assert* command in this case maintains that the condition must be true, otherwise it will *report* on the simulators console that the parity bit is incorrect and since the *severity* is failure, the simulation will stop. Alternatively, simulation will continue and "*Data received*" will be displayed on the console.

The mouse checks the command received by the host and reacts accordingly. Since our mouse controller only supports sending a 0xF4 to enable data reporting, our mouse test bench will check only for and handle that command. Since the 0xF4 was shifted in least-significant bit first, the data in the mouse register *m_data_in* is 0x2F. In this case, the mouse must send an acknowledge byte, 0xFA, followed by continuous data packets consisting of three 11-bit data transmissions. The first 11-bit data transmission contains the mouse status byte. The second and third contain *x*-movement and *y*-movement information, respectively. In our test bench, we have created some test data and send 2 data packets.

By setting the top level to be *mouse_top_tb* and running a simulation, a waveform is generated from the test bench. A segment of this waveform is shown in Fig. 7.8. This waveform shows the mouse releasing the *PS2C* clock after transmitting the device ID. At this point, the mouse is waiting for the host to pull the clock and data lines low at which time the mouse will begin generating a clock on *PS2C* again, shifting in the data in a host-to-device transmission.

Mouse

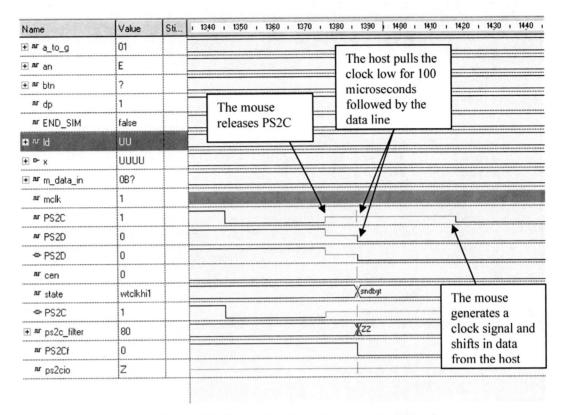

Figure 7.8 Simulation from the mouse test bench

Listing 7.7 mouse_top_tb.vhd

```vhdl
library ieee;
use work.ps2_components.all;
use ieee.std_logic_1164.all;

entity mouse_top_tb is
end mouse_top_tb;

architecture TB_ARCHITECTURE of mouse_top_tb is
    -- Component declaration of the tested unit
    component mouse_top
            port(
            mclk : in STD_LOGIC;
            PS2C: inout STD_LOGIC;
            PS2D: inout STD_LOGIC;
            btn : in STD_LOGIC_VECTOR(3 downto 0);
            ld : out STD_LOGIC_VECTOR(3 downto 0);
            a_to_g : out STD_LOGIC_VECTOR(6 downto 0);
            dp : out STD_LOGIC;
            an : out STD_LOGIC_VECTOR(3 downto 0)
        );
    end component;
```

Listing 7.7 (cont.) mouse_top_tb.vhd

```vhdl
        -- Stimulus signals
        -- signals mapped to the input and inout ports of tested entity
        signal mclk : std_logic;
        signal btn : std_logic_vector(3 downto 0);
        signal PS2C : std_logic;
        signal PS2D : std_logic;
        -- Observed signals
        -- signals mapped to the output ports of tested entity
        signal ld : std_logic_vector(7 downto 0);
        signal a_to_g : std_logic_vector(6 downto 0);
        signal dp : std_logic;
        signal an : std_logic_vector(3 downto 0);
        --8 bits of data, 1 parity, and 1 stop bit
        signal m_data_in : std_logic_vector(10 downto 0);
        --Signal is used to stop clock signal generators
        signal END_SIM: BOOLEAN:=FALSE;

begin
-- Unit Under Test port map
UUT : mouse_top
        port map (mclk => mclk, PS2C => PS2C, PS2D => PS2D, btn => btn,
                  ld => ld, a_to_g => a_to_g, dp => dp, an => an);

        --Stimulus Statements
mclk_tb: process
begin
        --Generate 50MHz clock
        --this process was generated based on formula:
        --    0 0 ns, 1 10 ns -r 20 ns
        --wait for <time to next event>; -- <current time>
        if END_SIM = FALSE then
              mclk <= '0';
              wait for 10 ns; --0 fs
        else
              wait;
        end if;
        if END_SIM = FALSE then
              mclk <= '1';
              wait for 10 ns; --10 ns
        else
              wait;
        end if;
end process;

tb: process
variable cnt:integer;
begin

        --Clear the design
        btn(3) <= '1';
```

Listing 7.7 (cont.) mouse_top_tb.vhd

```vhdl
    --Device-to-Host Communication Protocol
    --The mouse uses a protocol with 11-bit frames. These bits are:
    --  1 start bit. This is always 0.
    --    8 data bits, least significant bit first.
    --    1 parity bit (odd parity).
    --    1 stop bit. This is always 1.
    --    The mouse writes a bit on the data line when clock is high
    --    The host reads it when clock is low.

    --When the mouse powers on, it enters reset mode,
    --performs a self-test
    --then sends a X"AA" to the host.
    --Following the self-test success (X"AA"),
    --  it sends a device ID of X"00"
    wait for 1 us;   --self-test

    --Use mouse clock with frequency of 16kHz
    -- (62.5us period, 31.25 us between switch)

    PS2C <= '1';
    PS2D <= '0';        --start bit
    for x in 1 to 10 loop
         wait for 31.25 us;
         PS2C <= not PS2C; --0
         wait for 31.25 us;
         PS2C <= not PS2C; --1
         --Send "AA" LSB first "01010101"
         case x is
              when 1 => PS2D <= '0';
              when 2 => PS2D <= '1';
              when 3 => PS2D <= '0';
              when 4 => PS2D <= '1';
              when 5 => PS2D <= '0';
              when 6 => PS2D <= '1';
              when 7 => PS2D <= '0';
              when 8 => PS2D <= '1';
              when 9 => PS2D <= '0'; --Send Parity bit (3 1s)
              when 10 => PS2D <= '1'; --Send Stop bit
         end case;
    end loop;
    wait for 31.25 us;
    PS2C <= not PS2C; --0
    wait for 31.25 us;
    PS2C <= not PS2C; --1
    --End of Device-to-Host Communication X"AA"

    --Start of Device-to-Host Communication X"00" device ID
    PS2C <= '1';
    PS2D <= '0';        --start bit
    for x in 1 to 10 loop
         wait for 31.25 us;
         PS2C <= not PS2C; --0
         wait for 31.25 us;
         PS2C <= not PS2C; --1
```

Listing 7.7 (cont.) mouse_top_tb.vhd

```vhdl
                --Send "00" LSB first "00000000"
                case x is
                    when 1 => PS2D <= '0';
                    when 2 => PS2D <= '0';
                    when 3 => PS2D <= '0';
                    when 4 => PS2D <= '0';
                    when 5 => PS2D <= '0';
                    when 6 => PS2D <= '0';
                    when 7 => PS2D <= '0';
                    when 8 => PS2D <= '0';
                    when 9 => PS2D <= '1'; --Send Parity bit (0 1s)
                    when 10 => PS2D <= '1'; --Send Stop bit
                end case;
        end loop;
        wait for 31.25 us;
        PS2C <= not PS2C; --0
        wait for 31.25 us;
        PS2C <= not PS2C; --1
        --End of Device-to-Host Communication X"00" device ID
        PS2C <= 'Z';
        PS2D <= 'Z';

        wait for 10 us;
        btn(3) <= '0';
        --Host-to-Device Communication
        --The HOST will:
        -- 1  Bring the Clock line low for at least 100 microseconds.
        -- 2  Bring the Data line low.
        -- 3  Release the Clock line.
        -- 4  Wait for the device to bring the Clock line low.
        -- 5  Set/reset the Data line to send the first data bit
        -- 6  Wait for the device to bring Clock high.
        -- 7  Wait for the device to bring Clock low.
        -- 8  Repeat steps 5-7 for the other seven data bits
        --       and the parity bit
        -- 9  Release the Data line.
        -- 10 Wait for the device to bring Data low.
        -- 11 Wait for the device to bring Clock low.
        -- 12 Wait for the device to release Data and Clock

        wait until PS2C = '0';   --wait until clock line is held
                                 --low by host
        wait until PS2D = '0';   --wait until data line is low
        for x in 10 downto 0 loop
            wait for 31.25 us;
            PS2C <= '0';         --bring clock line low
            wait for 31.25 us;
            PS2C <= '1';         --bring clock line high
            m_data_in(x) <= PS2D;  --read data
        end loop;
        --after stop bit, device gives one last clock pulse
        PS2D <= '0';         --active low acknowledge
        wait for 31.25 us;
        PS2C <= '0';         --bring clock line low
        wait for 31.25 us;
```

Listing 7.7 (cont.) mouse_top_tb.vhd

```vhdl
        PS2C <= '1';        --bring clock line high
        PS2D <= '1';
    wait for 31.25 us;
        PS2C <= '0';
    wait for 31.25 us;

    --check parity bit, m_data_in(1)

    cnt := 0;
    for x in 9 downto 1 loop
            if m_data_in(x) = '1' then
                    cnt := cnt + 1;
            end if;
    end loop;

    assert (cnt = 1 or cnt = 3 or cnt = 5 or cnt = 7 or cnt = 9)
    report "Parity bit incorrect"
    severity failure;

    report "Data received";


    if m_data_in(9 downto 2) = X"2F" then
            --enter streaming mode
            --send back X"FA" acknowledge

            --Start of Device-to-Host Communication X"FA" acknowledge
            PS2C <= '1';
            PS2D <= '0';        --start bit
            for x in 1 to 10 loop
                wait for 31.25 us;
                PS2C <= not PS2C;       --0
                wait for 31.25 us;
                PS2C <= not PS2C;       --1
                --Send "FA" LSB first "01011111"
                case x is
                    when 1 => PS2D <= '0';
                    when 2 => PS2D <= '1';
                    when 3 => PS2D <= '0';
                    when 4 => PS2D <= '1';
                    when 5 => PS2D <= '1';
                    when 6 => PS2D <= '1';
                    when 7 => PS2D <= '1';
                    when 8 => PS2D <= '1';
                    when 9 => PS2D <= '1';  --Send Parity bit (0 1s)
                    when 10 => PS2D <= '1'; --Send Stop bit
                end case;
            end loop;
            wait for 31.25 us;
            PS2C <= not PS2C; --0
            wait for 31.25 us;
            PS2C <= not PS2C; --1
            --End of Device-to-Host Communication X"FA" acknowledge
```

Listing 7.7 (cont.) mouse_top_tb.vhd

```vhdl
            --generate some velocity vectors

            --Send Mouse Status Byte, X-movement Byte, Y-Movement Byte
            --Status Byte:  YOVF, XOVF, YSGN, XSGN, 1, MIDBTN,
            --   RIGHTBTN, LEFTBTN
            --Movement Bytes - two's complement numbers if negative

            -- left button, movement in x only
            -- 00001001 00101100 00000000
            --  Status     X         Y

         for x in 1 to 2 loop
            --do this twice... to send 2 sets of packets for testing.
    --Start of Device-to-Host Communication of Mouse Status Byte
            PS2C <= '1';
            PS2D <= '0';    --start bit
            for x in 1 to 10 loop
                wait for 31.25 us;
                PS2C <= not PS2C; --0
                wait for 31.25 us;
                PS2C <= not PS2C; --1
                --Send "09" LSB first "10010000"
                case x is
                    when 1 => PS2D <= '1';
                    when 2 => PS2D <= '0';
                    when 3 => PS2D <= '0';
                    when 4 => PS2D <= '1';
                    when 5 => PS2D <= '0';
                    when 6 => PS2D <= '0';
                    when 7 => PS2D <= '0';
                    when 8 => PS2D <= '0';
                    when 9 => PS2D <= '1'; --Send Parity bit (0 1s)
                    when 10 => PS2D <= '1'; --Send Stop bit
                end case;
            end loop;
            wait for 31.25 us;
            PS2C <= not PS2C;      --0
            wait for 31.25 us;
            PS2C <= not PS2C;      --1
       --End of Device-to-Host Communication of Mouse Status Byte

       --Start of Device-to-Host Communication of X-movement
            PS2C <= '1';
            PS2D <= '0';    --start bit
            for x in 1 to 10 loop
                wait for 31.25 us;
                PS2C <= not PS2C; --0
                wait for 31.25 us;
                PS2C <= not PS2C; --1
                --Send "2C" LSB first "00110100"
                case x is
                    when 1 => PS2D <= '0';
                    when 2 => PS2D <= '0';
                    when 3 => PS2D <= '1';
                    when 4 => PS2D <= '1';
```

Listing 7.7 (cont.) mouse_top_tb.vhd

```vhdl
                    when 5 => PS2D <= '0';
                    when 6 => PS2D <= '1';
                    when 7 => PS2D <= '0';
                    when 8 => PS2D <= '0';
                    when 9 => PS2D <= '1'; --Send Parity bit (0 1s)
                    when 10 => PS2D <= '1'; --Send Stop bit
                end case;
            end loop;
            wait for 31.25 us;
            PS2C <= not PS2C;      --0
            wait for 31.25 us;
            PS2C <= not PS2C;      --1
            --End of Device-to-Host Communication of X-movement

            --Start of Device-to-Host Communication of Y-movement
            PS2C <= '1';
            PS2D <= '0';   --start bit
            for x in 1 to 10 loop
                wait for 31.25 us;
                PS2C <= not PS2C; --0
                wait for 31.25 us;
                PS2C <= not PS2C; --1
                --Send "00" LSB first "00000000"
                case x is
                    when 1 => PS2D <= '0';
                    when 2 => PS2D <= '0';
                    when 3 => PS2D <= '0';
                    when 4 => PS2D <= '0';
                    when 5 => PS2D <= '0';
                    when 6 => PS2D <= '0';
                    when 7 => PS2D <= '0';
                    when 8 => PS2D <= '0';
                    when 9 => PS2D <= '1'; --Send Parity bit (0 1s)
                    when 10 => PS2D <= '1'; --Send Stop bit
                end case;
            end loop;
            wait for 31.25 us;
            PS2C <= not PS2C;      --0
            wait for 31.25 us;
            PS2C <= not PS2C;      --1
            --End of Device-to-Host Communication of Y-movement

        end loop;
        wait for 10us;
        report "END OF SIMULATION";
    else
        --wait for another command from mouse
        wait until PS2C'event and PS2C = '0';
        --wait until clock line is held low by host
        wait until PS2D <= '0'; --wait until data line is low
        --THEN PROCESS HOST DATA
--(since we are expecting an F4 and to enter streaming mode,
-- we will not process other host data in this testbench.)
    end if;
```

Listing 7.7 (cont.) mouse_top_tb.vhd
```
        END_SIM <= TRUE;

end process;

end TB_ARCHITECTURE;

configuration TESTBENCH_FOR_mouse_top of mouse_top_tb is
    for TB_ARCHITECTURE
        for UUT : mouse_top
            use entity work.mouse_top(mouse_top);
        end for;
    end for;
end TESTBENCH_FOR_mouse_top;
```

8

Graphics

In Example 39 in Chapter 6 we saw how the external RAM can be used as a video RAM that continuously refreshes the VGA screen. Each pixel on the screen corresponds to a particular color byte in memory as shown in Fig. 6.22. In Example 39 we simply read the contents of the RAM and displayed the loon photo on the screen. We can dynamically change the screen image by writing color bytes to the RAM. For example, filling the video RAM with zeros will clear the screen. We will do this in Example 43. In Example 44 we will show how to plot a dot at a particular (x,y) location on the screen by writing a single color byte to the corresponding RAM location.

Plotting a straight line between two arbitrary points on the screen can be done by using the Bresenham algorithm. We will explain and implement this algorithm in Example 45. In Example 46 we will illustrate how to plot a series of arbitrary straight lines by plotting a star. Finally, in Example 47 we will develop and implement an algorithm for plotting a circle on the screen.

Example 43

Clearing the Screen

The top-level design shown in Fig. 6.22 of Example 39 was used to continuously read the contents of the external RAM and display the color data on the VGA screen. We can modify this top-level design as shown in Fig. 8.1, which will allow us to write zeros to the RAM and thus clear the screen.

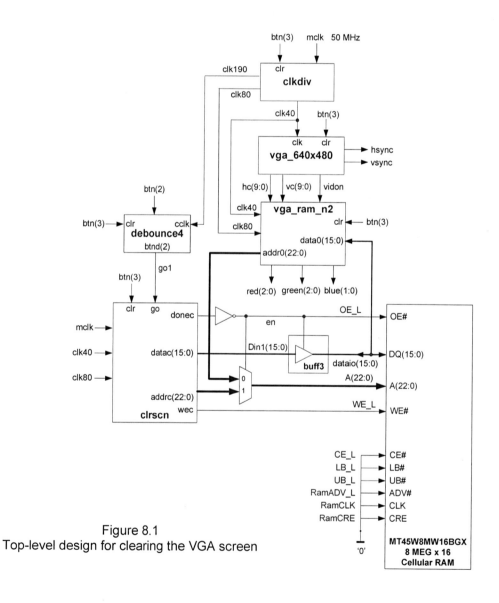

Figure 8.1
Top-level design for clearing the VGA screen

In Fig. 8.1 when *btn*(2) is pressed the *clrscn* component will make *donec* zero, which will enable the tri-state buffer, *buff3*, and allow *addrc*(22:0) to be connected to the RAM address line *A*(22:0). The output *datac*(15:0) from *clrscn* will be set to zero and the addresses *addrc*(22:0) will count from 0 to 153,600, writing 16 bits of zeros to each of these addresses, thus clearing the screen.

Recall that it will take 80 ns to write a word to the RAM. To clear the screen as quickly as possible we would like to increment *addrc*(22:0) every 80 ns. But the write enable signal *wec* must go low for at least 45 ns during this 80 ns period and then go high for a short time to latch the data into the RAM. We can generate such a *wec* output by ANDing together three clock signals with periods 20 ns, 40 ns, and 80 ns using the following statement.

```
wec <= (mclk and clk40 and clk80) or done_out;
```

This will result in the *wec* waveform shown in Fig. 8.2. Note that *wec* goes high for 10 ns each time that *addrc*(22:0) changes and then *wec* goes low for 70 ns. The *wec* output will also be high when *done_out* is high – i.e., when *donec* is high and control returns to the *vga_ram_n2* component in Fig. 8.1, which will continuously refresh the VGA screen.

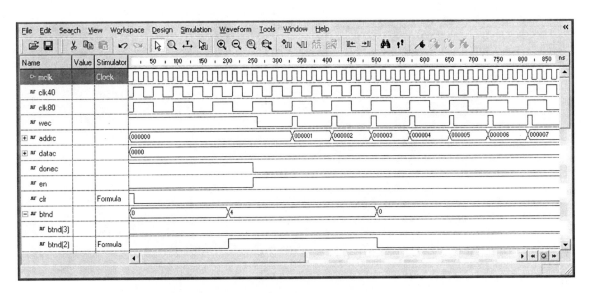

Figure 8.2 The output *wec* goes low for 70 ns every 80 ns.

Fig. 8.3 shows the state diagram that is implemented by the *clrscn* component. It waits in the *start* state until *go* goes high (i.e., until *btn*(2) is pressed). It will then stay in the *clear* state until all 153,600 words (*addr_max*) have been cleared to zero. It then goes to the *wtngo* state and stays there until *btn*(2) has been released.

Listing 8.1 shows the VHDL program for *clrscn*. The VHDL program for the top-level design in Fig. 8.1 is given in Listing 8.2.

Example 43

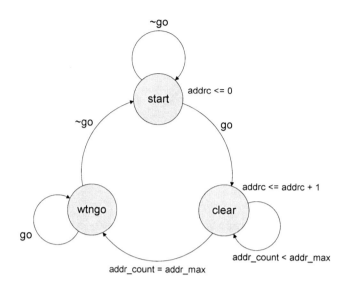

Figure 8.3 State diagram for clearing the screen

Listing 8.1 clrscn.vhd

```vhdl
-- Example 43a: clear screen
library IEEE;
use IEEE.STD_LOGIC_1164.ALL;
use IEEE.STD_LOGIC_UNSIGNED.ALL;

entity clrscn is
    port ( clk40 : in std_logic;
           clk80 : in std_logic;
           mclk : in std_logic;
           clr : in std_logic;
           go : in std_logic;
           donec : out std_logic;
           wec : out std_logic;
           addrc : out std_logic_vector(22 downto 0);
           datac : out std_logic_vector(15 downto 0)
    );
end clrscn;

architecture clrscn of clrscn is

type state_type is (start, clear, wtngo);
signal state: state_type;
signal done_out: std_logic;
signal addr_count: STD_LOGIC_VECTOR (19 downto 0);
constant addr_max: std_logic_vector(19 downto 0) := X"25800";
   --Max address = 320x480 = 153,600 = X"25800"

begin
datac <= X"0000";
wec <= (mclk and clk40 and clk80) or done_out;
donec <= done_out;
```

Listing 8.1 (cont.) clrscn.vhd

```vhdl
sreg: process(clk80, clr)
begin
  if clr = '1' then
    state <= start;
    addr_count <= (others => '0');
    done_out <= '1';
  elsif clk80'event and clk80 = '1' then
      case state is
        when start =>
            addr_count <= (others => '0');
              if go = '1' then
                  done_out <= '0';
                  state <= clear;
              else
                  done_out <= '1';
                  state <= start;
              end if;
        when clear =>
              if addr_count >= addr_max then
                  addr_count <= (others => '0');
                  state <= wtngo;
                  done_out <= '1';
              else
                  addr_count <= addr_count + 1;
                  done_out <= '0';
                    state <= clear;         -- stay in clear
              end if;
        when wtngo =>
              if go = '1' then
                  state <= wtngo;
              else
                  done_out <= '1';
                  state <= start;
              end if;
      end case;
  end if;
end process;

addrc <= "000" & addr_count;

end clrscn;
```

Listing 8.2 clrscn_top.vhd

```vhdl
-- Example 43b: clrscn_top
library IEEE;
use IEEE.STD_LOGIC_1164.all;
use IEEE.std_logic_unsigned.all;
use work.vga_components.all;

entity clrscn_top is
      port(
              mclk : in STD_LOGIC;
              btn : in STD_LOGIC_VECTOR(3 downto 0);
              sw : in STD_LOGIC_VECTOR(7 downto 0);
              hsync : out STD_LOGIC;
              vsync : out STD_LOGIC;
              red : out std_logic_vector(2 downto 0);
              green : out std_logic_vector(2 downto 0);
              blue : out std_logic_vector(1 downto 0);
              A : out STD_LOGIC_VECTOR(22 downto 0);
              DQ : inout STD_LOGIC_VECTOR(15 downto 0);
              CE_L : out STD_LOGIC;
              UB_L : out STD_LOGIC;
              LB_L : out STD_LOGIC;
              WE_L : out STD_LOGIC;
              OE_L : out STD_LOGIC;
              FlashCE_L : out STD_LOGIC;
              RamCLK : out STD_LOGIC;
              RamADV_L : out STD_LOGIC;
              RamCRE : out STD_LOGIC
           );
end clrscn_top;

architecture clrscn_top of clrscn_top is

      signal clr, clk40, clk80, clk190: std_logic;
      signal dataio, datac: std_logic_vector(15 downto 0);
      signal Din1: std_logic_vector(15 downto 0);
      signal wec, vidon, donec, en: std_logic;
      signal btnd: std_logic_vector(3 downto 0);
      signal hc, vc: std_logic_vector(9 downto 0);
      signal data0: std_logic_vector(15 downto 0);
      signal addr0, addrc: std_logic_vector(22 downto 0);
      begin

      clr <= btn(3);
      FlashCE_L <= '1';          -- Disable Flash
      CE_L <= '0';               -- Enable ram
      UB_L <= '0';
      LB_L <= '0';
      RamCLK <= '0';
      RamADV_L <= '0';
      RamCRE <= '0';
      WE_L <= wec;
      OE_L <= en;
      DQ <= dataio;
      en <= (not donec);
      Din1 <= datac;
```

Listing 8.2 (cont.) clrscn_top.vhd

```vhdl
U1 : clkdiv3
     port map(mclk => mclk, clr => clr, clk190 => clk190,
              clk80 => clk80, clk40 => clk40);

U2 : buff3
     generic map(N => 16)
     port map(input => Din1, en => en, output => dataio);

U3 : vga_ram_n2
     port map(clk40 => clk40, clk80 => clk80, clr => clr,
              vidon => vidon, hc => hc, vc => vc,
              data0 => dataio, addr0 => addr0, red => red,
              green => green, blue => blue);

U4 : clrscn
     port map(clk40 => clk40, clk80 => clk80,
              mclk => mclk, clr => clr, go => btnd(2),
              done => donec, wec => wec, addrc => addrc,
              datac => datac);

U5 : debounce4
     port map(cclk => clk190, clr => clr, inp => btn,
              outp => btnd);

U6 : mux2g
     generic map(N => 23)
     port map(a => addr0, b => addrc, s => en, y => A);

U7 : vga_640x480
     port map(clk => clk40, clr => clr, hsync => hsync,
              vsync => vsync, hc => hc, vc => vc,
              vidon => vidon);

end clrscn_top;
```

Example 44

Plotting a Dot

To plot a dot at location (*x,y*) on the screen you would need to read the word from the external RAM at the address corresponding to the pixel at (*x,y*) as shown in Fig. 6.22 of Example 39. You would then replace the proper (upper or lower) pixel byte in the word with the color byte of the dot to be plotted, and re-write the word to the RAM. This would plot the dot without changing the color of the adjacent pixel.

Fig. 8.4 shows the additions to the top-level design in Fig. 8.1 that will allow you to plot a dot whose location is determined by the switches on the Nexys-2 board when you press *btn*(0). The *x*(9:0) and *y*(9:0) locations are given by

```
x <= '0' & sw(7 downto 4) & "00001";
y <= '0' & sw(3 downto 0) & "00001";
```

so that the dot can be plotted anywhere on the screen with resolution of about 32 pixels.

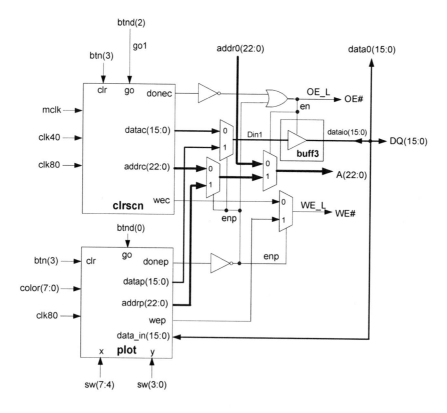

Figure 8.4 Additions to the top-level design in Fig. 8.1 to plot a dot

Fig. 8.5 shows the state diagram that is implemented in the component *plot*. After waiting for *btn*(0) to be pressed (*go* = '1') this state diagram computes the RAM address *addrp*(22:0), reads the word at that address, sets the color pixel of the dot to be plotted, writes the word back to the RAM, and then waits for *btn*(0) to be released. Listing 8.3 is the VHDL program that implements this *plot* component. Listing 8.4 is the VHDL program for the top-level design shown in Figs. 8.1 and 8.4.

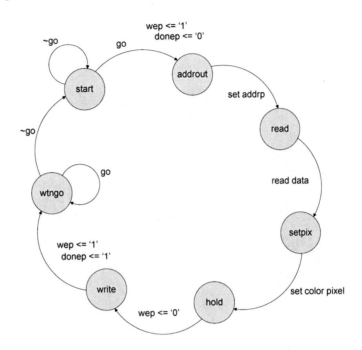

Figure 8.5 State diagram for plotting a dot

Listing 8.3 plot.vhd

```
-- Example 44a: Plot a dot at (x,y)
library IEEE;
use IEEE.STD_LOGIC_1164.ALL;
use IEEE.STD_LOGIC_UNSIGNED.ALL;

entity plot is
    port ( clk80 : in std_logic;
           clr : in std_logic;
           go : in std_logic;
           x, y : in std_logic_vector(9 downto 0);
           data_in : in std_logic_vector(15 downto 0);
           color : in std_logic_vector(7 downto 0);
           donep : out std_logic;
           wep : out std_logic;
           addrp : out std_logic_vector(22 downto 0);
           datap : out std_logic_vector(15 downto 0)
           );
end plot;
```

Listing 8.3 (cont.) plot.vhd

```vhdl
architecture plot of plot is

type state_type is (start, addrout, read, setpix, hold,
                                            write, wtngo);
signal state: state_type;
signal ram_addr: STD_LOGIC_VECTOR (22 downto 0);
signal data: STD_LOGIC_VECTOR (15 downto 0);
signal px: STD_LOGIC;


begin

   process(x, y)
   variable ram_addr1: STD_LOGIC_VECTOR (22 downto 0);
   variable ram_addr2: STD_LOGIC_VECTOR (22 downto 0);
   begin
         ram_addr1 := ("00000" & y & "00000000") +
                     ("0000000" & y & "000000");  -- y*(256+64)
         ram_addr2 := ram_addr1 +
                ("00000000000000" & x(9 downto 1)); -- y*320+x/2
         px <= x(0);
         ram_addr <= ram_addr2;
   end process;

sreg: process(clk80, clr)
begin
  if clr = '1' then
         state <= start;
         addrp <= (others => '0');
         wep <= '1';
         donep <= '1';
  elsif clk80'event and clk80 = '1' then
    case state is
      when start =>
          wep <= '1';
          if go = '1' then
              state <= addrout;
              donep <= '0';
          else
              state <= start;
          end if;
      when addrout =>
          state <= read;
          addrp <= ram_addr;
          wep <= '1';
      when read =>
          state <= setpix;
          data <= data_in;
          wep <= '1';
```

Listing 8.3 (cont.) plot.vhd

```vhdl
            when setpix =>
                state <= hold;
                wep <= '1';
                if px = '0' then
                    datap <= color & data(7 downto 0);
                else
                    datap <= data(15 downto 8) & color;
                end if;
            when hold =>
                state <= write;
                wep <= '0';
            when write =>
                state <= wtngo;
                donep <= '1';
                wep <= '1';
            when wtngo =>
                wep <= '1';
                if go = '1' then
                    state <= wtngo;
                else
                    state <= start;
                end if;
            when others =>
                null;
        end case;
      end if;
    end process;

end plot;
```

Listing 8.4 plotdot_top.vhd

```vhdl
-- Example 44b: plotdot_top
library IEEE;
use IEEE.STD_LOGIC_1164.all;
use IEEE.std_logic_unsigned.all;
use work.vga_components.all;

entity plotdot_top is
    port(
            mclk : in STD_LOGIC;
            btn : in STD_LOGIC_VECTOR(3 downto 0);
            sw : in STD_LOGIC_VECTOR(7 downto 0);
            hsync : out STD_LOGIC;
            vsync : out STD_LOGIC;
            red : out std_logic_vector(2 downto 0);
            green : out std_logic_vector(2 downto 0);
            blue : out std_logic_vector(1 downto 0);
            A : out STD_LOGIC_VECTOR(22 downto 0);
            DQ : inout STD_LOGIC_VECTOR(15 downto 0);
            CE_L : out STD_LOGIC;
            UB_L : out STD_LOGIC;
            LB_L : out STD_LOGIC;
            WE_L : out STD_LOGIC;
            OE_L : out STD_LOGIC;
            FlashCE_L : out STD_LOGIC;
            RamCLK : out STD_LOGIC;
            RamADV_L : out STD_LOGIC;
            RamCRE : out STD_LOGIC
         );
end plotdot_top;

architecture plotdot_top of plotdot_top is
    signal clr, clk40, clk80, clk190: std_logic;
    signal donec, donep, en, enp: std_logic;
    signal dataio, datac, datap: std_logic_vector(15 downto 0);
    signal wec, wep, vidon: std_logic;
    signal btnd: std_logic_vector(3 downto 0);
    signal hc, vc, x, y: std_logic_vector(9 downto 0);
    signal color_dot: std_logic_vector(7 downto 0);
    signal data0, Din1: std_logic_vector(15 downto 0);
    signal addr0,addr1,addrc,addrp: std_logic_vector(22 downto 0);
    begin
        clr <= btn(3);
        FlashCE_L <= '1';      -- Disable Flash
        CE_L <= '0';           -- Enable ram
        UB_L <= '0';
        LB_L <= '0';
        RamCLK <= '0';
        RamADV_L <= '0';
        RamCRE <= '0';
        OE_L <= en;    -- enable data bus
        DQ <= dataio;
        en <= (not donec) or (not donep);
        enp <= not donep;
```

Listing 8.4 (cont.) plotdot_top.vhd

```vhdl
            y <= '0' & sw(3 downto 0) & "00001";
            x <= '0' & sw(7 downto 4) & "00001";
            color_dot <= "11111111";  -- white dot

U1 : clkdiv3
   port map(mclk => mclk, clr => clr, clk190 => clk190,
            clk80 => clk80, clk40 => clk40);

U2 : buff3
   generic map(N => 16)
   port map(input => Din1, en => en, output => dataio);

U3 : vga_ram_n2
   port map(clk40 => clk40, clk80 => clk80, clr => clr,
            vidon => vidon, hc => hc, vc => vc,
            data0 => dataio, addr0 => addr0, red => red,
            green => green, blue => blue);

U4 : clrscn
   port map(clk40 => clk40, clk80 => clk80,
            mclk => mclk, clr => clr, go => btnd(2),
            done => donec, wec => wec, addrc => addrc,
            datac => datac);

U5 : debounce4
   port map(
            cclk => clk190, clr => clr, inp => btn, outp => btnd);

U6 : mux2g
   generic map(N => 23)
   port map(a => addr0, b => addr1, s => en, y => A);

U7 : vga_640x480
   port map(clk => clk40, clr => clr, hsync => hsync,
            vsync => vsync, hc => hc, vc => vc, vidon => vidon);

U8 : plot
   port map(clk80 => clk80, clr => clr, go => btnd(0),
            x => x, y => y, data_in => dataio,
            color => color_dot, donep => donep, wep => wep,
            addrp => addrp, datap => datap);

U9 : mux2g
   generic map(N => 23)
   port map(a => addrc, b => addrp, s => enp, y => addr1);

U10 : mux2g
   generic map(N => 16)
   port map(a => datac, b => datap, s => enp, y => Din1);

U11 : mux2
   port map(a => wec, b => wep, s => enp, y => WE_L);

end plotdot_top;
```

Example 45

Plotting a Line

In Example 44 we saw how to plot a dot anywhere on the screen. To draw a straight line between any two dots you need to figure out which dots to plot to make a straight line. There are several algorithms that you could use to do this. One of the more popular ones is called Bresenham's line algorithm. We will first describe this algorithm and then implement it in VHDL.

Bresenham's Line Algorithm

In this section we will describe the Bresenham algorithm for plotting a straight line. More details about this alogorithm can be found on the web.[1] Fig. 8.6 shows which pixels you would plot to draw a straight line from $(x0, y0)$ to $(x1, y1)$ as best you can on a video screen made up of discrete pixels. In this figure we assume that the line slopes downward to the right at an angle that is less than 45 degrees. The slope of the line is given by

$$\text{slope } m = \frac{deltay}{deltax} \qquad (8\text{-}1)$$

where $deltax = \text{abs}(x1 - x0)$ and $deltay = \text{abs}(y1 - y0)$.

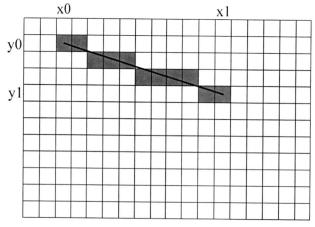

Figure 8.6 Plotting a line on a video screen

---

[1] http://en.wikipedia.org/wiki/Bresenham's_line_algorithm
   http://www.cs.unc.edu/~mcmillan/comp136/Lecture6/Lines.html

Note in Fig. 8.6 that we plot a dot for every value of x between $x0$ and $x1$. The only question is whether to plot the dot at the current value of y or at $y + 1$. To see which choice we should make we have redrawn a portion of Fig. 8.6 in greater detail in Fig. 8.7 where we show only six pixels with the dots plotted in the center of the pixel. We start with a dot plotted at (x, y), which is exactly on the line we wish to plot. The next dot will be plotted at $(x+1, y)$ because the *error*, defined as

$$\text{slope } m = \frac{deltay}{deltax} = error$$

is lesss than 0.5. Had the error been greater than 0.5 then we would have plotted the dot at $(x+1, y+1)$. Note in Fig. 8.8 that the errors accumulate and, in general, is given by $error = error + m$.

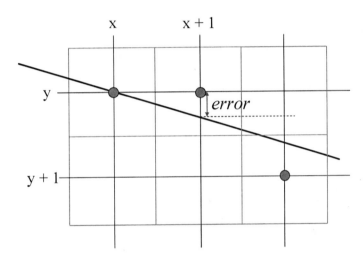

Figure 8.7 Defining the error in plotting a dot

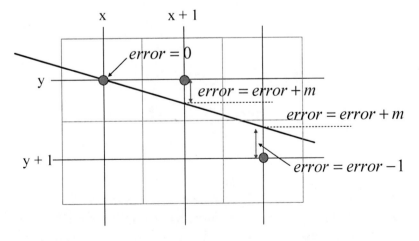

Figure 8.8 Plot the dot at (x+1, y+1) when $error \geq 0.5$

From Fig. 8.8 we see that when the *error* gets greater than 0.5 we plot the dot at $(x+1, y+1)$ and subtract 1 from the *error*. The basic algorithm shown in Listing 8.5 will plot a line from $(x0, y0)$ to $(x1, y1)$ according to Figs. 8.6 – 8.8. The first problem in implementing this algorithm on an FPGA is that *error* and *m* are *real* variables. Note that $m \times deltax = deltay$. If we multiply both sides of equations in Listing 8.5 by *deltax*, and call $error \times deltax = eps$, then we obtain the integer algorithm shown in Listing 8.6.

Listing 8.5 Basic algorithm for plotting dots in Fig. 8.8
```
plotline(xo,y0,x1,y1)
        int deltax = abs(x1-x0);
        int deltay = abs(y1-y0);
        real error = 0;
        real m = deltay/deltax;
        int y = y0;
        int x = x0;
        while(x <= x1){
          plot(x,y);
          error = error + m;
          if(error >= 0.5){
            y = y + 1;
            error = error - 1;
          )
          x++;
        }
```

Listing 8.6 Integer algorithm for plotting dots in Fig. 8.8
```
plotline(xo,y0,x1,y1)
        int deltax = abs(x1-x0);
        int deltay = abs(y1-y0);
        int eps = 0;
        int y = y0;
        int x = x0;
        while(x <= x1){
          plot(x,y);
          eps = eps + deltay;
          if(2*eps >= deltax){
            y = y + 1;
            eps = eps - deltax;
          )
          x++;
        }
```

The integer algorithm in Listing 8.6 only works when the line to be plotted slopes downward to the right at an angle less than or equal to 45 degrees as shown in Fig. 8.6 – 8.8. Suppose the line has a steeper slope, say from (2,5) to (9,18), as shown in Fig. 8.9. Here we can use a trick by noting that if we swap *x0* and *y0* and also swap *x1* and *y1* we obtain the line from (5,2) to (18,9), which has a slope less than 45 degrees, and therefore the algorithm in Listing 8.6 will work. To get the real steep line plotted we only need to remember to *plot(y,x)* rather than *plot(x,y)*. We will therefore add the following code to our algorithm:

```
boolean steep = abs(y1 - y0) > abs(x1 - x0)
if(steep){
    swap(x0,y0);
    swap(x1,y1);
}
-----
if(steep){
    plot(y,x);
else
    plot(x,y);
}
```

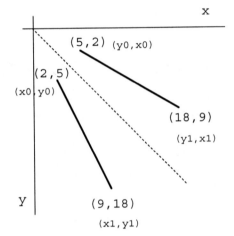

Figure 8.9 Plotting a line with a steep slope

The only other case we need to consider is when the line slopes upward to the right as shown in Fig. 8.10. In this case if we swap *x0* and *x1* and also swap *y0* and *y1* we can plot the line using our current algorithm if we just remember to plot dots at $(x+1, y-1)$ rather than at $(x+1, y+1)$. We will therefore add the following code to our algorithm:

```
if(x0 > x1){
    swap(x0,x1);
    swap(y0,y1);
}

if(y0 < y1){
    ystep = 1;
else
    ystep = -1;
}
```

The complete algorithm for plotting a line from (*x0*, *y0*) to (*x1*, *y1*) is given in Listing 8.7. We will now look at how to implement this algorithm in VHDL.

Example 45

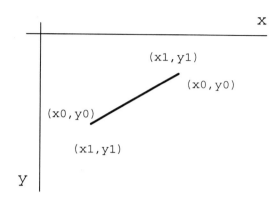

Figure 8.10 Plotting a line sloping upward to the right

Listing 8.7 Complete algorithm for plotting a line from (x0,y0) to (x1,y1)
```
plotline(x0,y0,x1,y1)
    int deltax = abs(x1-x0);
    int deltay = abs(y1-y0);
    int eps = 0;
    int y = y0;
    int x = x0;
    boolean steep = abs(y1 - y0) > abs(x1 - x0)
    if(steep){
        swap(x0,y0);
        swap(x1,y1);
    }
    if(x0 > x1){
        swap(x0,x1);
        swap(y0,y1);
    }
    if(y0 < y1){ ystep = 1; else ystep = -1;}
    while(x <= x1){
      if(steep){ plot(y,x); else plot(x,y); }
      eps = eps + deltay;
      if(2*eps >= deltax){
          y = y + ystep;
          eps = eps - deltax;
      )
      x++;
    }
```

Plotting a Line in VHDL

When implementing the algorithm shown in Listing 8.7 in VHDL the first thing to examine is what parts of the algorithm can be implemented as combinational logic and what parts need to be sequential. Note in Listing 8.7 that all of the code before the *while* statement can be implemented as combinational logic in which the inputs are (*x*0, *y*0, *x*1, *y*1) and the outputs are possibly shuffled versions of (*x*0, *y*0, *x*1, *y*1) together with values for *deltax*, *deltay*, *steep*, and *ystep*. If a microprocessor were used to implement this algorithm separate instructions, each taking at least one clock cycle, would be used to implement each of these setup instructions. In our FPGA, by contrast, these outputs are generated combinationally and therefore take zero clock cycles.

The *while* loop portion of the line plotting algorithm in Listing 8.7 will be implemented using the state diagram shown in Fig. 8.11. Listing 8.8 is the VHDL program that implements the entire line plotting algorithm. The *setup* process implements the combinational logic described above. The *wloop* process implements the *while* loop in Listing 8.7 following the state diagram in Fig. 8.11. The *plotchk* process in Listing 8.8 swaps *x* and *y* for plotting depending on the value of *steep*.

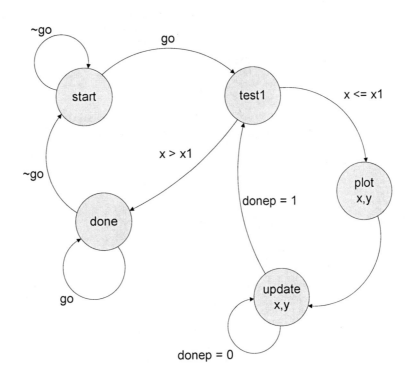

Figure 8.11 State diagram for plotting a line

Fig. 8.12 shows the additions to the top-level design in Fig. 8.4 that we will use to plot lines on the VGA screen. Using this top-level design we can plot dots in the same way we did in Example 44 using the switches. The last two dots plotted are remembered in the registers *x*0, *y*0, *x*1, and *y*1. Pressing *btn*(1) will then plot a line between these two dots. You can continue in this fashion plotting a series of lines connected together. The VHDL program for this top-level design is given in Listing 8.9. A sample run of this line-drawing program is shown in Fig. 8.13.

Listing 8.8 plotline.vhd

```vhdl
--   Example 45a: plotline
library IEEE;
use IEEE.STD_LOGIC_1164.ALL;
use IEEE.STD_LOGIC_ARITH.ALL;
use IEEE.STD_LOGIC_UNSIGNED.ALL;

entity plotline is
    port(
         clk80 : in STD_LOGIC;
         clr : in STD_LOGIC;
         go : in STD_LOGIC;
         x0 : in STD_LOGIC_VECTOR(9 downto 0);
         y0 : in STD_LOGIC_VECTOR(9 downto 0);
         x1 : in STD_LOGIC_VECTOR(9 downto 0);
         y1 : in STD_LOGIC_VECTOR(9 downto 0);
         donep : in STD_LOGIC;
         color : in STD_LOGIC_VECTOR(7 downto 0);
         donepl : out STD_LOGIC;
         goplot : out STD_LOGIC;
         color_dot : out STD_LOGIC_VECTOR(7 downto 0);
         x : out STD_LOGIC_VECTOR(9 downto 0);
         y : out STD_LOGIC_VECTOR(9 downto 0)
        );
end plotline;


architecture plotline of plotline is
type state_type is (start,test1,plot,update,done);
signal state: state_type;
signal deltax, deltay, ystep, eps: STD_LOGIC_VECTOR(9 downto 0);
signal xs, ys, x0s, y0s, x1s, y1s,: STD_LOGIC_VECTOR(9 downto 0);
signal steep : STD_LOGIC;
begin

   color_dot <= color;

setup: process(x0, y0, x1, y1)
variable x0v,y0v,x1v,y1v,xm,ym: STD_LOGIC_VECTOR(9 downto 0);
variable steepv: STD_LOGIC;
variable delxv, delyv, temp: STD_LOGIC_VECTOR(9 downto 0);
begin
   x0v := x0;
   x1v := x1;
   y0v := y0;
   y1v := y1;
   xm := x1v - x0v;
   ym := y1v - y0v;
   if xm(9) = '1' then
        xm := 0 - xm;
   end if;
   if ym(9) = '1' then
        ym := 0 - ym;
   end if;
```

Listing 8.8 (cont.) plotline.vhd

```vhdl
        if ym > xm then
                steepv := '1';
        else
                steepv := '0';
        end if;
        if steepv = '1' then
                temp := x0v;            -- swap(x0,y0)
                x0v := y0v;
                y0v := temp;
                temp := x1v;            -- swap(x1,y1)
                x1v := y1v;
                y1v := temp;
        end if;
        if x0v > x1v then
                temp := x0v;            -- swap(x0,x1)
                x0v := x1v;
                x1v := temp;
                temp := y0v;            -- swap(y0,y1)
                y0v := y1v;
                y1v := temp;
        end if;
        if y0v < y1v then
                ystep <= "0000000001";  -- 1
        else
                ystep <= "1111111111";  -- -1
        end if;
        xm := x1v - x0v;
        ym := y1v - y0v;
        if xm(9) = '1' then
                xm := 0 - xm;
        end if;
        if ym(9) = '1' then
                ym := 0 - ym;
        end if;
        deltax <= xm;
        deltay <= ym;
        x0s <= x0v;
        y0s <= y0v;
        x1s <= x1v;
        y1s <= y1v;
        steep <= steepv;
end process setup;

wloop: process(clk80, clr)
variable xv, yv: STD_LOGIC_VECTOR(9 downto 0);
variable epsv, delxv, delyv: signed(9 downto 0);
begin
   for i in 0 to 9 loop
           delxv(i) := deltax(i);
           delyv(i) := deltay(i);
   end loop;
```

Listing 8.8 (cont.) plotline.vhd

```vhdl
    if clr = '1' then
       state <= start;
       goplot <= '0';
       donepl <= '1';
       xs <= x0s;
       ys <= y0s;
    elsif clk80'event and clk80 = '1' then
      case state is
        when start =>
           goplot <= '0';
           donepl <= '1';
           epsv := "0000000000";
           xs <= x0s;
           ys <= y0s;
           if go = '1' then
              donepl <= '0';
              state <= test1;
           else
              state <= start;
           end if;
        when test1 =>
           if xs <= x1s then
              state <= plot;
              goplot <= '1';
           else
              state <= done;
           end if;
        when plot =>
           goplot <= '0';
           epsv := epsv + delyv;
           state <= update;
        when update =>
           if donep = '0' then
              state <= update;
           else
              if (epsv(8 downto 0) & '0') >= delxv then
                 ys <= ys + ystep;
                 epsv := epsv - delxv;
              end if;
              xs <= xs + 1;
              state <= test1;
           end if;
        when done =>
           donepl <= '1';
           if go = '1' then
              state <= done;
            else
               state <= start;
            end if;
        when others =>
           null;
      end case;
   end if;
```

Listing 8.8 (cont.) plotline.vhd

```vhdl
      for i in 0 to 9 loop
            eps(i) <= epsv(i);
      end loop;
end process wloop;

plotchk: process(xs, ys, steep)
variable xv, yv, temp: STD_LOGIC_VECTOR(9 downto 0);
begin
    xv := xs;
    yv := ys;
    if steep = '1' then
            temp := xv;        -- swap(x0,y0)
            xv := yv;
            yv := temp;
    end if;
    x <= xv;
    y <= yv;
end process plotchk;

end plotline;
```

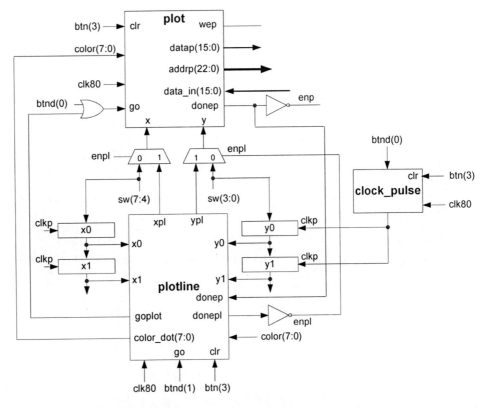

Figure 8.12 Additions to the top-level design in Fig. 8.4 to plot a line

Listing 8.9 plotline_top.vhd

```vhdl
-- Example 45b: plotline_top
library IEEE;
use IEEE.STD_LOGIC_1164.all;
use IEEE.std_logic_unsigned.all;
use work.vga_components.all;

entity plotline_top is
    port(
            mclk : in STD_LOGIC;
            btn : in STD_LOGIC_VECTOR(3 downto 0);
            sw : in STD_LOGIC_VECTOR(7 downto 0);
            hsync : out STD_LOGIC;
            vsync : out STD_LOGIC;
            red : out std_logic_vector(2 downto 0);
            green : out std_logic_vector(2 downto 0);
            blue : out std_logic_vector(1 downto 0);
            A : out STD_LOGIC_VECTOR(22 downto 0);
            DQ : inout STD_LOGIC_VECTOR(15 downto 0);
            CE_L : out STD_LOGIC;
            UB_L : out STD_LOGIC;
            LB_L : out STD_LOGIC;
            WE_L : out STD_LOGIC;
            OE_L : out STD_LOGIC;
            FlashCE_L : out STD_LOGIC;
            RamCLK : out STD_LOGIC;
            RamADV_L : out STD_LOGIC;
            RamCRE : out STD_LOGIC
        );
end plotline_top;

architecture plotline_top of plotline_top is

    signal clr, clk40, clk80, clk190, clkp: std_logic;
    signal donec, donep, donepl, en, enp, enpl: std_logic;
    signal dataio, datac, datap: std_logic_vector(15 downto 0);
    signal wec, wep, vidon, one, goplot, goplotl: std_logic;
    signal btnd: std_logic_vector(3 downto 0);
    signal hc,vc,xpl,ypl,xsw,ysw: std_logic_vector(9 downto 0);
    signal x, y, x0, y0, x1, y1: std_logic_vector(9 downto 0);
    signal color, color_dot: std_logic_vector(7 downto 0);
    signal data0, Din1: std_logic_vector(15 downto 0);
    signal addr0,addr1,addrc,addrp: std_logic_vector(22 downto 0);
    begin

    clr <= btn(3);
    FlashCE_L <= '1';          -- Disable Flash
    CE_L <= '0';               -- Enable ram
    UB_L <= '0';
    LB_L <= '0';
    RamCLK <= '0';
    RamADV_L <= '0';
    RamCRE <= '0';
    OE_L <= en;                -- enable data bus
    DQ <= dataio;
```

Listing 8.9 (cont.) plotline_top.vhd

```vhdl
      en <= (not donec) or (not donep);
      enp <= not donep;
      enpl <= not donepl;
      ysw <= '0' & SW(3 downto 0) & "00001";
      xsw <= '0' & SW(7 downto 4) & "00001";
      color <= "11111111";  -- white dots and line
      one <= '1';
      goplot <= btnd(0) or goplotl;

   U1 : clkdiv3
      port map(mclk => mclk, clr => clr, clk190 => clk190,
            clk80 => clk80, clk40 => clk40);

   U2 : buff3
      generic map(N => 16)
      port map(input => Din1, en => en, output => dataio);

   U3 : vga_ram_n2
      port map(clk40 => clk40, clk80 => clk80, clr => clr,
            vidon => vidon, hc => hc, vc => vc, data0 => dataio,
            addr0 => addr0, red => red, green => green,
            blue => blue);

   U4 : clrscn
      port map(clk40 => clk40, clk80 => clk80, mclk => mclk,
            clr => clr, go => btnd(2), done => donec, wec => wec,
            addrc => addrc, datac => datac);

   U5 : debounce4
      port map(cclk => clk190, clr => clr, inp => btn,
            outp => btnd);

   U6 : mux2g
      generic map(N => 23)
      port map(a => addr0, b => addr1, s => en, y => A);


   U7 : vga_640x480
      port map(clk => clk40, clr => clr, hsync => hsync,
            vsync => vsync, hc => hc, vc => vc, vidon => vidon);

   U8 : plot
      port map(clk80 => clk80, clr => clr, go => goplot, x => x,
            y => y, data_in => dataio, color => color_dot,
            donep => donep, wep => wep, addrp => addrp,
            datap => datap);

   U9 : mux2g
      generic map(N => 23)
      port map(a => addrc, b => addrp, s => enp, y => addr1);

   U10 : mux2g
      generic map(N => 16)
      port map(a => datac, b => datap, s => enp, y => Din1);
```

Listing 8.9 (cont.) plotline_top.vhd

```vhdl
U11 : mux2
   port map(a => wec, b => wep, s => enp, y => WE_L);


U12 : plotline
   port map(clk80 => clk80, clr => clr, go => btnd(1), x0 => x0,
           y0 => y0, x1 => x1, y1 => y1, donep => donep,
           color => color, donepl => donepl, goplot => goplotl,
           color_dot => color_dot, x => xpl, y => ypl);

x0reg : reg
   generic map(N => 10)
   port map(d => xsw, load => one, clr => clr, clk => clkp,
           q => x0);

x1reg : reg
   generic map(N => 10)
   port map(d => x0, load => one, clr => clr, clk => clkp,
           q => x1);

y0reg : reg
   generic map(N => 10)
   port map(d => ysw, load => one, clr => clr, clk => clkp,
           q => y0);

y1reg : reg
   generic map(N => 10)
   port map(d => y0, load => one, clr => clr, clk => clkp,
           q => y1);

U13 : clock_pulse
   port map(inp => btnd(0), cclk => clk80, clr => clr,
           outp => clkp);

U14 : mux2g
   generic map(N => 10)
   port map(a => xsw, b => xpl, s => enpl, y => x);

U15 : mux2g
   generic map(N => 10)
   port map(a => ysw, b => ypl, s => enpl, y => y);

end plotline_top;
```

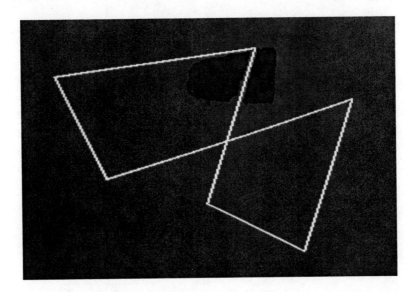

Figure 8.13 Sample run of the top-level design in Fig. 8.12 and Listing 8.9

Example 46

Plotting a Star

In Example 45 we used the switches to set the end points of lines to draw on the screen. In this example we will show how to store a list of end points in a ROM so that the figure can be drawn all at once. As an example we will draw the star with the end point coordinates shown in Fig. 8.14.

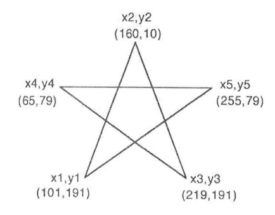

Figure 8.14 Defining the screen coordinates of a star

We will use a VHDL ROM on the type described in Example 27 to store the star coordinates. Each word in the ROM will be 20 bits wide with the upper 10 bits containing the *x*-coordinate of a point and the lower 10 bits containing the *y*-coordinate of a point. Listing 8.10 gives the VHDL program for this ROM that contains the star coordinates shown in Fig. 8.14. Note that the ROM has a 3-bit address and we have included all eight words – with the last three words containing the coordinates of (*x*1,*y*1).

Fig. 8.15 shows the modifications to the line plotting top-level design in Fig. 8.12 that we will make to plot the star. We must read the first two words from the ROM and store the contents in the registers (*x*0,*y*0) and (*x*1,*y*1). After plotting that line the next word in the ROM is read into (*x*1,*y*1) and the old values from (*x*1,*y*1) are shifted into (*x*0,*y*0). Thus, the next line to be plotted will be the line from (*x*2,*y*2) to (*x*3,*y*3) in Fig. 8.14. This process will continue until all five lines are plotted.

The state diagram that will implement this process when *btn*(0) is pressed is shown in Fig. 8.16. This state diagram is implemented in the *plot_star* component in Fig. 8.15 and Listing 8.11 is the VHDL program that implements this *plot_star* component. The VHDL program for the top-level design in Fig. 8.15 is given in Listing 8.12. A sample run of this star plotting program is shown in Fig. 8.17.

Listing 8.10 prom_star.vhd

```vhdl
-- Example 46a: prom_star
library IEEE;
use IEEE.std_logic_1164.all;
use IEEE.std_logic_unsigned.all;

entity prom_star is
    port (
        addr: in STD_LOGIC_VECTOR (2 downto 0);
        M: out STD_LOGIC_VECTOR (19 downto 0)
    );
end prom_star;

architecture prom_star of prom_star is
type rom_array is array (NATURAL range <>) of
                                STD_LOGIC_VECTOR (19 downto 0);
constant rom: rom_array := (
"00010000000010100000",      -- 64, 160
"00100000000000110000",      -- 128, 48
"00110000000010100000",      -- 192, 160
"00001100000001010000",      -- 48, 80
"00110100000001010000",      -- 208, 80
"00010000000010100000",      -- 64, 160
"00010000000010100000",      -- 64, 160
"00010000000010100000");     -- 64, 160
begin
  process(addr)
  variable j: integer;
  begin
    j := conv_integer(addr);
    M <= rom(j);
  end process;
end prom_star;
```

Example 46

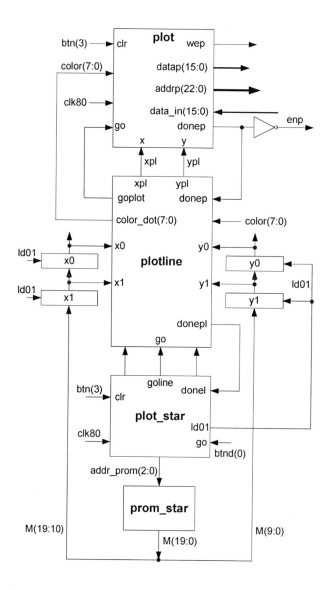

Figure 8.15 Modifications of the top-level design in Fig. 8.12 to plot a star

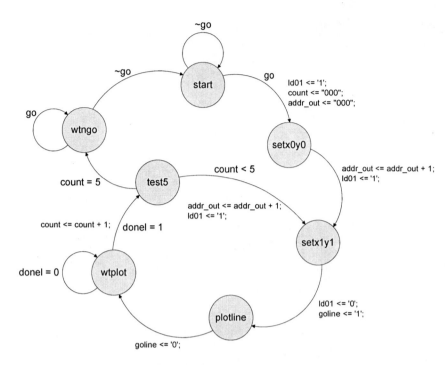

Figure 8.16 State diagram implemented in the component *plot_star* in Fig. 8.15

Listing 8.11 plot_star.vhd

```
--    Example 46b: plot_star
library IEEE;
use IEEE.STD_LOGIC_1164.ALL;
use IEEE.STD_LOGIC_UNSIGNED.ALL;

entity plot_star is
    port ( clk80 : in std_logic;
           clr : in std_logic;
           go : in std_logic;
           donel : in std_logic;
             addr_prom : out STD_LOGIC_VECTOR(2 downto 0);
             ld01: out STD_LOGIC;
             goline: out STD_LOGIC
    );
end plot_star;

architecture plot_star of plot_star is
type state_type is
(start,setx0y0,setx1y1,plotline,wtplot,test5,wtngo);
signal state: state_type;
signal count, addr_out: STD_LOGIC_VECTOR(2 downto 0);

begin
addr_prom <= addr_out;

sreg: process(clk80, clr)
begin
```

Listing 8.11 (cont.) plot_star.vhd

```vhdl
   if clr = '1' then
        state <= start;
        goline <= '0';
        ld01   <= '0';
        count <= "000";
        addr_out <= "000";
   elsif clk80'event and clk80 = '1' then
     case state is
       when start =>
            if go = '1' then
                    state <= setx0y0;
                    ld01 <= '1';
                    count <= "000";
                    addr_out <= "000";
            else
                    state <= start;
            end if;
       when setx0y0 =>
            state <= setx1y1;
            addr_out <= addr_out + 1;
            ld01 <= '1';
       when setx1y1 =>
            state <= plotline;
            ld01 <= '0';
            goline <= '1';
       when plotline =>
            state <= wtplot;
            goline <= '0';
       when wtplot =>
            if done1 = '1' then
                    state <= test5;
                    count <= count + 1;
            else
                    state <= wtplot;
            end if;
       when test5 =>
            if count = "101" then
                    state <= wtngo;
            else
                    state <= setx1y1;
                    addr_out <= addr_out + 1;
                    ld01 <= '1';
            end if;
       when wtngo =>
            if go = '1' then
                    state <= wtngo;
            else
                    state <= start;
            end if;
       when others =>
            null;
     end case;
   end if;
end process;
end plot_star;
```

Listing 8.12 plot_star_top.vhd

```vhdl
-- Example 46c: plot_star_top
library IEEE;
use IEEE.STD_LOGIC_1164.all;
use IEEE.std_logic_unsigned.all;
use work.vga_components.all;

entity plot_star_top is
    port(
        mclk : in STD_LOGIC;
        btn : in STD_LOGIC_VECTOR(3 downto 0);
        hsync : out STD_LOGIC;
        vsync : out STD_LOGIC;
        red : out std_logic_vector(2 downto 0);
        green : out std_logic_vector(2 downto 0);
        blue : out std_logic_vector(1 downto 0);
        A : out STD_LOGIC_VECTOR(22 downto 0);
        DQ : inout STD_LOGIC_VECTOR(15 downto 0);
        CE_L : out STD_LOGIC;
        UB_L : out STD_LOGIC;
        LB_L : out STD_LOGIC;
        WE_L : out STD_LOGIC;
        OE_L : out STD_LOGIC;
        FlashCE_L : out STD_LOGIC;
        RamCLK : out STD_LOGIC;
        RamADV_L : out STD_LOGIC;
        RamCRE : out STD_LOGIC
        );
end plot_star_top;

architecture plot_star_top of plot_star_top is
    signal clr, clk40, clk80, clk190: std_logic;
    signal donec, donep, donepl, en, enp, ld01: std_logic;
    signal dataio, datac, datap: std_logic_vector(15 downto 0);
    signal wec, wep, vidon, one, goplot, goline: std_logic;
    signal btnd: std_logic_vector(3 downto 0);
    signal hc, vc, x, y, xpl, ypl: std_logic_vector(9 downto 0);
    signal x0, y0, x1, y1: std_logic_vector(9 downto 0);
    signal color, color_dot: std_logic_vector(7 downto 0);
    signal data0, Din1: std_logic_vector(15 downto 0);
    signal addr0,addr1,addrc,addrp: std_logic_vector(22 downto 0);
    signal M: std_logic_vector(19 downto 0);
    signal addr_prom: std_logic_vector(2 downto 0);
    begin

        clr <= btn(3);
        FlashCE_L <= '1';           -- Disable Flash
        CE_L <= '0';                -- Enable ram
        UB_L <= '0';
        LB_L <= '0';
        RamCLK <= '0';
        RamADV_L <= '0';
        RamCRE <= '0';
        OE_L <= en;                 -- enable data bus
        DQ <= dataio;
```

Listing 8.12 (cont.) plot_star_top.vhd

```vhdl
         en <= (not donec) or (not donep);
         enp <= not donep;
         color <= "11111111";    -- white dots and line
         one <= '1';

   U1 : clkdiv3
         port map(mclk => mclk, clr => clr, clk190 => clk190,
                clk80 => clk80, clk40 => clk40);

   U2 : buff3
         generic map(N => 16)
         port map(input => Din1, en => en, output => dataio);

   U3 : vga_ram_n2
         port map(clk40 => clk40, clk80 => clk80, clr => clr,
                vidon => vidon, hc => hc, vc => vc, data0 => dataio,
                addr0 => addr0, red => red, green => green, blue => blue);

   U4 : clrscn
         port map(clk40 => clk40, clk80 => clk80, mclk => mclk,
                clr => clr, go => btnd(2), done => donec, wec => wec,
                addrc => addrc, datac => datac);

   U5 : debounce4
         port map(cclk => clk190, clr => clr, inp => btn, outp => btnd);

   U6 : mux2g
         generic map(N => 23)
         port map(a => addr0, b => addr1, s => en, y => A);

   U7 : vga_640x480
         port map(
                clk => clk40, clr => clr, hsync => hsync, vsync => vsync,
                hc => hc, vc => vc, vidon => vidon);

   U8 : plot
         port map(clk80 => clk80, clr => clr, go => goplot, x => xpl,
                y => ypl, data_in => dataio, color => color_dot,
                donep => donep, wep => wep, addrp => addrp, datap => datap);

   U9 : mux2g
         generic map(N => 23)
         port map(a => addrc, b => addrp, s => enp, y => addr1);

   U10 : mux2g
         generic map(N => 16)
         port map(a => datac, b => datap, s => enp, y => Din1);

   U11 : mux2
         port map(a => wec, b => wep, s => enp, y => WE_L);
```

Listing 8.12 (cont.) plot_star_top.vhd

```
U12 : plotline
    port map(clk80 => clk80, clr => clr, go => goline, x0 => x0,
             y0 => y0, x1 => x1, y1 => y1, donep => donep,
             color => color, donepl => donepl, goplot => goplot,
             color_dot => color_dot, x => xpl, y => ypl);

x1reg : reg
    generic map(N => 10)
    port map(d => M(19 downto 10), load => ld01, clr => clr,
             clk => clk80, q => x1);

x0reg : reg
    generic map(N => 10)
    port map(d => x1, load => ld01, clr => clr, clk => clk80,
             q => x0);

y1reg : reg
    generic map(N => 10)
    port map(d => M(9 downto 0), load => ld01, clr => clr,
             clk => clk80, q => y1);

y0reg : reg
    generic map(N => 10)
    port map(d => y1, load => ld01, clr => clr, clk => clk80,
             q => y0);

U13 : plot_star
    port map(clk80 => clk80, clr => clr, go => btnd(0),
             donel => donepl, addr_prom => addr_prom, ld01 => ld01,
             goline => goline);

U14 : prom_star
    port map(addr => addr_prom, M => M);

end plot_star_top;
```

Figure 8.17 Sample run of the top-level design in Fig. 8.15 and Listing 8.12

Example 47

Plotting a Circle

In Example 45 we saw how we could use Bresenham's line algorithm to plot lines on the VGA screen. In this example we will develop a similar Bresenham algorithm for plotting a circle and implement it in VHDL. You can find information on such algorithms on the web.[2]

A Circle Plotting Algorithm

Our goal is to plot a circle of radius r centered at (xc,yc) on the VGA screen as shown in Fig. 8.18 where we have introduced a local u-v coordinate system with the origin at the center of the circle. A point on the circle at location (u,v) will have the screen coordinates

$$x = xc + u \qquad y = yc - v \qquad (8\text{-}2)$$

The equation of the circle is

$$u^2 + v^2 = r^2 \qquad (8\text{-}3)$$

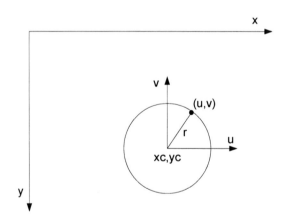

Figure 8.18 Geometry used to plot a circle

How can we determine what dots to plot to make a circle? It turns out that the symmetry of the circle will simplify things. Given xc, yc, and r we can plot a dot at

---

[2] e.g., http://www.cs.unc.edu/~mcmillan/comp136/Lecture7/circle.html

$$(u,v) = (0,r) = (xc, yc - r) \qquad (8\text{-}4)$$

By symmetry, we can immediately plot the additional three dots at

$$(u,v) = (0,-r) = (xc, yc + r) \qquad (8\text{-}5)$$
$$(u,v) = (r,0) = (xc + r, yc) \qquad (8\text{-}6)$$
$$(u,v) = (-r,0) = (xc - r, yc) \qquad (8\text{-}7)$$

as shown in Fig. 8.19a.

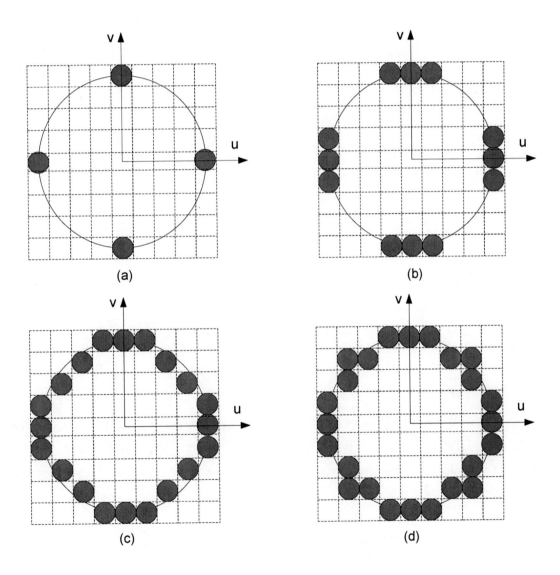

Figure 8.19 Steps used to plot a circle

We will now start at $(u,v) = (0,r)$ and move around the circle clockwise. At $u = 1$ we would plot the dot at either $v = r$ or $v = r - 1$ depending on how much the circle sloped down. From Fig. 8.19a the circle has moved down less than 0.5 and therefore we should plot the next dot at $(u,v) = (1,r)$. But once we have plotted this dot we note that we have 8-way symmetry about the u-v axes as well as the $u = v$ axes at 45 degrees. We can therefore plot the eight additional dots shown in Fig. 8.19b. At this point we continue to increment u by 1 (u is now 2) and decide whether to plot the dot at the current v position ($v = r$) or at $v - 1$ ($v = r - 1$). From Fig. 8.19b it looks as if we should choose $v - 1$, and using the 8-way symmetry we can plot the additional eight dot as shown in Fig. 8.19c.

In Fig. 8.19 the value of r is 4 and at this point the value of v is $r - 1 = 3$. If we increment u again it will have a value of 3, which is equal to v. This value will be on the circle so we plot it at (u,v). By 4-way symmetry we also plot three additional dots at $(u,-v), (-u,v)$, and $(-u,-v)$ as shown in Fig. 8.19d. At this point the complete circle has been plotted. Note that starting at the top of the circle and moving clockwise, we needed to consider how to plot dots in only the first 1/8 of the circle.

In this first octant how do we determine whether to plot the next dot at $(u+1,v)$ or at $(u+1,v-1)$? This is similar to the problem we had in deciding what dot to plot next when plotting a straight line as shown in Fig. 8.7. To find the criteria for a circle we can rewrite the equation of the circle from Eq. (8-3) as

$$f(u,v) = u^2 + v^2 - r^2 = 0 \qquad (8\text{-}8)$$

The function $f(u,v)$ is called a *discriminating function*. This function will be equal to zero for all points on the circle. If a point (u,v) is outside the circle, then $f(u,v) > 0$, while if a point (u,v) is inside the circle, then $f(u,v) < 0$.

To see how we can use this discriminating function to decide where to plot the next dot consider the expanded circle segment shown in Fig. 8.20. After plotting the first dot at $u = 0$, $v = r$, the next dot will be plotted at either $(u+1,v)$ or $(u+1,v-1)$. If you plot the dot at $(u+1,v)$, then the dot will be outside the circle and $f(u+1,v) > 0$. On the other hand, if you plot the dot at $(u+1,v-1)$, then the dot will be inside the circle and $f(u+1,v-1) < 0$. From Fig. 8.20 it is clear that you should plot the dot at $(u+1,v)$ because the real circle is above the midpoint between v and $v - 1$. This suggests that we calculate the discriminating function at this midpoint, which will be

$$f(1, r - \tfrac{1}{2}) = 1^2 + \left(r - \tfrac{1}{2}\right)^2 - r^2 = \tfrac{5}{4} - r = \frac{5 - 4r}{4} \qquad (8\text{-}9)$$

Now if this value is less than zero as it is in Fig. 8.20, then we are inside the circle and the dot should be plotted at $(u+1,v)$ as shown in Fig. 8.20.

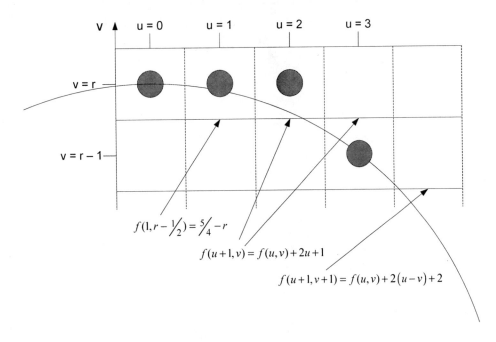

Figure 8.20 The discriminating function is defined at the midpoint between *v* and *v*-1

At this point we need to compute the discriminating function at the next midpoint at $u = 2$, $v = r - 1/2$, or, in general, at $(u+1, v)$. We can therefore write

$$f(u+1,v) = (u+1)^2 + v^2 - r^2 = u^2 + 2u + 1 + v^2 - r^2$$

or, using Eq. (8-8)

$$f(u+1,v) = f(u,v) + 2u + 1 \qquad (8\text{-}10)$$

Note that we can calculate the next value of $f(u+1,v)$ in terms of the current value of $f(u,v)$. In Fig. 8.20 this next value will be less than zero because we are still inside the circle and therefore the next dot should be plotted at $(u+1,v) = (2,r)$ as shown. However, when we calculate the next value of $f(u+1,v)$ using Eq. (8-10), its value will be greater than zero because the midpoint is now outside the circle, so we should plot the next dot at $(u+1,v-1) = (3, r-1)$ as shown in Fig. 8.20.

We now need to compute the discriminating function at the next midpoint at $(u+1, v-1)$. We can therefore write

$$f(u+1,v-1) = (u+1)^2 + (v-1)^2 - r^2 = u^2 + 2u + 1 + v^2 - 2v + 1 - r^2$$

or, using Eq. (8-8)

$$f(u+1,v-1) = f(u,v) + 2(u-v) + 2 \qquad (8\text{-}11)$$

If this value is greater than zero, as it will be in Fig. 8.20 because the midpoint is now outside the circle, we would plot the next dot at $(u+1, v-1)$. This process will continue for the first octant, i.e., until $u = v$, at which point the entire circle will have been plotted. This algorithm is described by Listing 8.13.

Listing 8.13 plot_circle algorithm
```
plot_circle(xc, yc, r)
{
    int u = 0;
    int v = r;
    int f = (5 - r*4)/4;
    plot4(xc, yc, r);
    while (u < v) {
        u = u + 1;
        if (f < 0)
            f = f + 2*u + 1;
        else {
            v = v - 1;
            f = f + 2*(u - v) + 1;
        }
        if (u = v)
            plot4a(xc, yc, u, v);
        else
            plot8(xc, yc, u, v);
    }
}
```

After calculating the initial value of *f* using Eq. (8-9) the four dots in Fig. 8.19a are plotted with the function *plot4(xc, yc, r)*. While still in the first octant ($u < v$) the algorithm increments *u* and if $f < 0$, *f* is updated using Eq. (8-10), otherwise *v* is decremented and *f* is updated using Eq. (8-11). If $u = v$, the octant is complete and the four dots at 45 degrees are plotted using the function *plot4a(xc, yc, u, v)*, otherwise the eight symmetric dots are plotted using the function *plot8(xc, yc, u, v)*.

Plotting a Circle in VHDL

Fig. 8.21 shows the state diagram used to implement the algorithm in Listing 8.13. Fig. 8.22 shows the additions we will make to Fig. 8.4 to plot a circle. We will use the switches to load the center of the circle into registers *xc* and *yc* and the radius of the circle into register *r*. Pressing *btn*(1) will then plot the circle.

In Fig. 8.22 the component *plot_circle* will implement the state diagram in Fig. 8.21. The VHDL program for the *plot_circle* component is given in Listing 8.14. The VHDL program for the top-level design in Fig. 8.21 is given in Listing 8.15. A sample run of this top-level design is shown in Fig. 8.23.

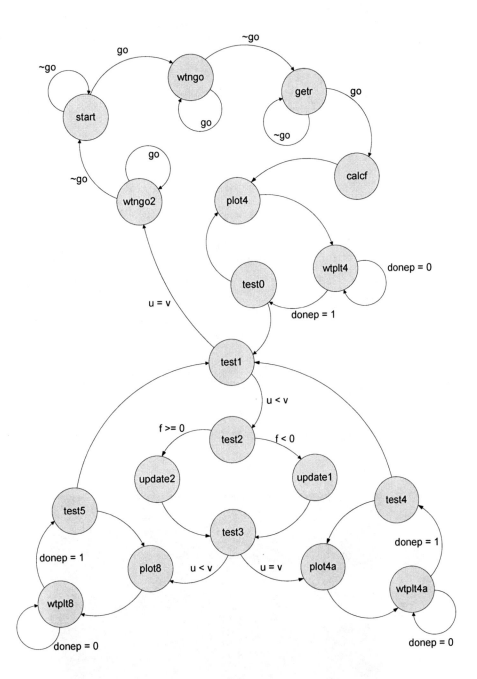

Figure 8.21 State diagram used to implement the *plot_circle* algorithm in Listing 8.13

Example 47

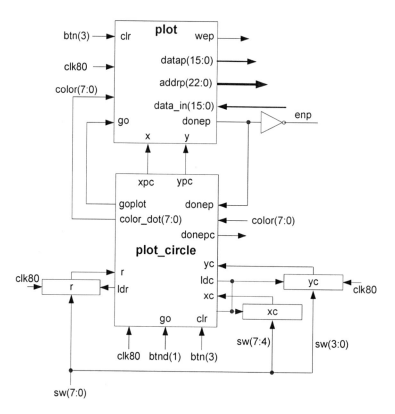

Figure 8.22 Additions to the top-level design in Fig. 8.4 to plot a circle

Figure 8.23 Sample run of the top-level design in Fig. 8.22 and Listing 8.15

Listing 8.14 plot_circle.vhd

```vhdl
--    Example 47a: plot_circle
library IEEE;
use IEEE.STD_LOGIC_1164.ALL;
use IEEE.STD_LOGIC_UNSIGNED.ALL;

entity plot_circle is
   port(
         clk80 : in STD_LOGIC;
         clr : in STD_LOGIC;
         go : in STD_LOGIC;
         xc : in STD_LOGIC_VECTOR(9 downto 0);
         yc : in STD_LOGIC_VECTOR(9 downto 0);
         r : in STD_LOGIC_VECTOR(9 downto 0);
         donep : in STD_LOGIC;
         color : in STD_LOGIC_VECTOR(7 downto 0);
         donepc : out STD_LOGIC;
         goplot : out STD_LOGIC;
         ldc, ldr : out STD_LOGIC;
         color_dot : out STD_LOGIC_VECTOR(7 downto 0);
         x : out STD_LOGIC_VECTOR(9 downto 0);
         y : out STD_LOGIC_VECTOR(9 downto 0)
        );
end plot_circle;


architecture plot_circle of plot_circle is
type state_type is (start,getr,calcf,plot4,wtplt4,test0,test1,
      test2,test3,test4,test5,update1,update2,plot4a,wtplt4a,
      plot8,wtplt8,wtngo,wtngo2);
signal state: state_type;
signal xs, ys, u, v, umv, f, f0: STD_LOGIC_VECTOR(9 downto 0);
signal count: std_logic_vector(3 downto 0);
begin

   color_dot <= color;
   x <= xs;
   y <= ys;
   f0 <= 5 - (r(7 downto 0) & "00");
   umv <= u - v;

csm: process(clk80, clr)
variable uv, vv: std_logic_vector(9 downto 0);
begin
  if clr = '1' then
     state <= start;
     goplot <= '0';
     donepc <= '1';
     f <= (others => '0');
     uv := (others => '0');
     vv := (others => '0');
     count <= "0000";
     ldc <= '0';
     ldr <= '0';
  elsif clk80'event and clk80 = '1' then
```

Listing 8.14 (cont.) plot_circle.vhd

```vhdl
      case state is
        when start =>
           goplot <= '0';
           donepc <= '1';
           if go = '1' then
                 uv := (others => '0');
                 vv := (others => '0');
                 count <= "0000";
                 ldc <= '1';          -- load xc, yc
                 state <= wtngo;
           else
                 state <= start;
           end if;
        when wtngo =>
           ldc <= '0';
           if go = '1' then
                 state <= wtngo;
           else
                 state <= getr;
           end if;
        when getr =>
           if go = '1' then
                 ldr <= '1';
                 state <= calcf;
           else
                 state <= getr;
           end if;
        when calcf =>
           ldr <= '0';
           donepc <= '0';
           goplot <= '1';
           f <= f0(9) & f0(9) & f0(9 downto 2);   -- f = (5-4*r)/4
           state <= plot4;
        when plot4 =>
           goplot <= '0';
           state <= wtplt4;
        when wtplt4 =>
           if donep = '0' then
                 state <= wtplt4;
           else
                 count <= count + 1;
                 state <= test0;
           end if;
        when test0 =>
           if count = "0100" then
                 state <= test1;
                 uv := uv + 1;
                 vv := r;
           else
                 goplot <= '1';
                 state <= plot4;
           end if;
```

Listing 8.14 (cont.) plot_circle.vhd

```vhdl
            when test1 =>
               if u < v then
                     state <= test2;
               else
                     state <= wtngo2;
               end if;
            when test2 =>
               if f(9) = '1' then              -- f < 0
                     state <= update1;
               else
                     vv := vv - 1;
                     state <= update2;
               end if;
            when update1 =>
               f <= f + (u(8 downto 0) & '0') + 1;    -- f = f+2*u+1
               state <= test3;
            when update2 =>
               f <= f+(umv(8 downto 0) & '0')+2;      -- f = f+2*(u-v)+2
               state <= test3;
            when test3 =>
               goplot <= '1';
               count <= "0100";
               if u = v then
                     state <= plot4a;
               else
                     state <= plot8;
               end if;
            when plot4a =>
               goplot <= '0';
               state <= wtplt4a;
            when wtplt4a =>
               if donep = '0' then
                     state <= wtplt4a;
               else
                     count <= count + 1;
                     state <= test4;
               end if;
            when test4 =>
               if count = "1000" then
                     uv := uv + 1;
                     state <= test1;
               else
                     goplot <= '1';
                     state <= plot4a;
               end if;
            when plot8 =>
               goplot <= '0';
               state <= wtplt8;
            when wtplt8 =>
               if donep = '0' then
                     state <= wtplt8;
               else
                     count <= count + 1;
                     state <= test5;
               end if;
```

Listing 8.14 (cont.) plot_circle.vhd

```vhdl
            when test5 =>
                if count = "1100" then
                    uv := uv + 1;
                    state <= test1;
                else
                    goplot <= '1';
                    state <= plot8;
                end if;
            when wtngo2 =>
                donepc <= '1';
                if go = '1' then
                    state <= wtngo2;
                else
                    state <= start;
                end if;
            when others =>
                null;
        end case;
    end if;
    u <= uv;
    v <= vv;
end process csm;

setxy: process(xc, yc, r, u, v, count)
begin
    case count is
        when "0000" =>
            xs <= xc;
            ys <= yc - r;
        when "0001" =>
            xs <= xc;
            ys <= yc + r;
        when "0010" =>
            xs <= xc + r;
            ys <= yc;
        when "0011" =>
            xs <= xc - r;
            ys <= yc;
        when "0100" =>
            xs <= xc + u;
            ys <= yc + v;
        when "0101" =>
            xs <= xc + u;
            ys <= yc - v;
        when "0110" =>
            xs <= xc - u;
            ys <= yc + v;
        when "0111" =>
            xs <= xc - u;
            ys <= yc - v;
        when "1000" =>
            xs <= xc + v;
            ys <= yc + u;
```

Listing 8.14 (cont.) plot_circle.vhd

```vhdl
      when "1001" =>
          xs <= xc + v;
          ys <= yc - u;
      when "1010" =>
          xs <= xc - v;
          ys <= yc + u;
      when "1011" =>
          xs <= xc - v;
          ys <= yc - u;
      when others =>
          null;
    end case;
end process setxy;

end plot_circle;
```

Listing 8.15 plot_circle_top.vhd

```vhdl
-- Example 47b: plot_circle_top
library IEEE;
use IEEE.STD_LOGIC_1164.all;
use IEEE.std_logic_unsigned.all;
use work.vga_components.all;

entity plot_circle_top is
    port(
            mclk : in STD_LOGIC;
            btn : in STD_LOGIC_VECTOR(3 downto 0);
            sw : in STD_LOGIC_VECTOR(7 downto 0);
            hsync : out STD_LOGIC;
            vsync : out STD_LOGIC;
            red : out std_logic_vector(2 downto 0);
            green : out std_logic_vector(2 downto 0);
            blue : out std_logic_vector(1 downto 0);
            A : out STD_LOGIC_VECTOR(22 downto 0);
            DQ : inout STD_LOGIC_VECTOR(15 downto 0);
            CE_L : out STD_LOGIC;
            UB_L : out STD_LOGIC;
            LB_L : out STD_LOGIC;
            WE_L : out STD_LOGIC;
            OE_L : out STD_LOGIC;
            FlashCE_L : out STD_LOGIC;
            RamCLK : out STD_LOGIC;
            RamADV_L : out STD_LOGIC;
            RamCRE : out STD_LOGIC
        );
end plot_circle_top;
```

Listing 8.15 (cont.) plot_circle_top.vhd

```vhdl
architecture plot_circle_top of plot_circle_top is
signal clr, clk40, clk80, clk190, ldc, ldr: std_logic;
signal donec, donep, donepc, en, enp: std_logic;
signal dataio, datac, datap: std_logic_vector(15 downto 0);
signal wec, wep, vidon, goplot: std_logic;
signal btnd: std_logic_vector(3 downto 0);
signal hc, vc, xpc, ypc, xsw, ysw: std_logic_vector(9 downto 0);
signal xc, yc, r: std_logic_vector(9 downto 0);
signal color, color_dot, r8: std_logic_vector(7 downto 0);
signal Din1: std_logic_vector(15 downto 0);
signal addr0, addr1, addrc, addrp: std_logic_vector(22 downto 0);
begin

   clr <= btn(3);
   FlashCE_L <= '1';        -- Disable Flash
   CE_L <= '0';             -- Enable ram
   UB_L <= '0';
   LB_L <= '0';
   RamCLK <= '0';
   RamADV_L <= '0';
   RamCRE <= '0';
   OE_L <= en;              -- enable data bus
   DQ <= dataio;
   en <= (not donec) or (not donep);
   enp <= not donep;
   ysw <= '0' & sw(3 downto 0) & "00001";
   xsw <= '0' & sw(7 downto 4) & "00001";
   r <= "00" & r8;
   color <= "11111111";     -- white dots and line

U1 : clkdiv3
   port map(mclk => mclk, clr => clr, clk190 => clk190,
            clk80 => clk80, clk40 => clk40);

U2 : buff3
   generic map(N => 16)
   port map(input => Din1, en => en, output => dataio);

U3 : vga_ram_n2
   port map(clk40 => clk40, clk80 => clk80, clr => clr,
            vidon => vidon, hc => hc, vc => vc, data0 => dataio,
            addr0 => addr0, red => red, green => green,
            blue => blue);

U4 : clrscn
   port map(clk40 => clk40, clk80 => clk80, mclk => mclk,
            clr => clr, go => btnd(2), done => donec, wec => wec,
            addrc => addrc, datac => datac);

U5 : debounce4
   port map(cclk => clk190, clr => clr, inp => btn,
            outp => btnd);
```

Listing 8.15 (cont.) plot_circle_top.vhd

```vhdl
U6 : mux2g
   generic map(N => 23)
   port map(a => addr0, b => addr1, s => en, y => A);

U7 : vga_640x480
   port map(clk => clk40, clr => clr, hsync => hsync,
         vsync => vsync, hc => hc, vc => vc, vidon => vidon);

U8 : plot
   port map(clk80 => clk80, clr => clr, go => goplot, x => xpc,
         y => ypc, data_in => dataio, color => color_dot,
         donep => donep, wep => wep, addrp => addrp,
         datap => datap);

U9 : mux2g
   generic map(N => 23)
   port map(a => addrc, b => addrp, s => enp, y => addr1);

U10 : mux2g
   generic map(N => 16)
   port map(a => datac, b => datap, s => enp, y => Din1);

U11 : mux2
   port map(a => wec, b => wep, s => enp, y => WE_L);

U12 : plot_circle
   port map(clk80 => clk80, clr => clr, go => btnd(1), xc => xc,
         yc => yc, r => r, donep => donep, color => color,
         donepc => donepc, goplot => goplot, ldc => ldc,
         ldr => ldr, color_dot => color_dot, x => xpc, y => ypc);

xcreg : reg
   generic map(N => 10)
   port map(d => xsw, load => ldc, clr => clr, clk => clk80,
         q => xc);

ycreg : reg
   generic map(N => 10)
   port map(d => ysw, load => ldc, clr => clr, clk => clk80,
         q => yc);

rreg : reg
   generic map(N => 8)
   port map(d => sw, load => ldr, clr => clr, clk => clk80,
         q => r8);

end plot_circle_top;
```

9

Forth Core for FPGAs

The VHDL programs that we have written so far in this book implement what we call *special-purpose processors* (SPPs). For example, Examples 23 and 25 produced special-purpose processors with the only job of computing the greatest common divisor. Similarly, Examples 24 and 26 produced special-purpose processors with the only job of calculating an integer square root. We created a UART SPP in Chapter 4 and a screen saver SPP in Example 38. The SPP in Example 46 plotted a star and the SPP in Example 47 plotted circles on the VGA screen.

Microprocessors are *general-purpose processors* (GPPs) in which a program (a set of instructions) is stored in memory, and then when the program is run each instruction is loaded into the processor and executed in turn. General-purpose processors have the advantage that when you want to solve a different problem you just have to change the program (software) and the same hardware will execute the new instructions. The disadvantage of a GPP is that it will always take more time to execute the software instructions than if you had designed a SPP using VHDL to solve the particular problem. This is because it takes time to fetch each instruction from memory, decode it, and execute it. In addition, as we have seen, when you design a SPP using VHDL you can often execute more than one equivalent software instruction at the same time.

GPPs are convenient, however, because it is often faster to just write another software program than to design completely new hardware. Of course, as we have seen in this book, designing new hardware involves writing software in the form of VHDL programs. If you get good at this, you may be able to design fast hardware SPPs as quickly as your software friends can design software programs that will run much slower on a GPP!

In fact, you can consider a GPP to be just another SPP, the purpose of which is to execute some specific instruction set. In this chapter we will show you how you can design a GPP core using VHDL and implement it on an FPGA. You will then be able to write high-level software programs, compile them to your microprocessor core that you designed, and run the software on your FPGA.

Most microprocessors execute some specific low-level instruction set. When you write a program in assembly language there is a one-to-one correspondence between the assembly language instructions and the microprocessor instruction set. Such programs are tedious to write and more commonly you would write a program in some high-level language, often C, which is then compiled to the corresponding assembly language instructions. One high-level C instruction will often require several assembly language instructions to implement.

In this chapter we choose to implement a microprocessor core that will execute high-level Forth programs directly – without having to go through any kind of assembly language. This will allow you to write high-level Forth programs and execute them on the Forth core that we design in this chapter. You are probably not familiar with the Forth programming language, so we will begin by giving you a brief primer on Forth.

The Forth Programming Language

Chuck Moore invented Forth in the late 1960s while programming minicomputers in assembly language. His idea was to create a simple system that would allow him to write many more useful programs than he could by using assembly language. The essence of Forth is simplicity -- always try to do things in the simplest possible way. Forth is a way of thinking about problems in a modular way. It is modular in the extreme. Everything in Forth is a word and every word is a module that does something useful. There is an action associated with Forth words. The words execute themselves. In this sense they are very object-oriented. Words are sent parameters on a data stack, the words are executed, and results are left on the data stack. We really don't care how the word does it -- once we have written it and tested it so we know that it works.

Forth uses Reverse Polish Notation (RPN) to evaluate arithmetic expressions. If you use a Hewlett-Packard calculator, you are already familiar with this notation. For example, if you want to calculate

$$(3+5)*(7-2)$$

you would write

$$3\ 5\ +\ 7\ 2\ -\ *$$

We use a 16-bit data stack of the type described in Example 29 to evaluate this expression. First 3 and 5 are pushed on the stack. Then + adds the top two elements on the stack and leaves the sum 8 on the stack. Next 7 and 2 are pushed on the stack and then – subtracts 2 from 7 and leaves the difference 5 on the stack. At this point 8 and 5 will be on the stack and * multiplies these two numbers and leaves the product 40 on the stack.

Forth programs require that we are able to manipulate the elements on the data stack. Table 9.1 is a list of the Forth data stack manipulation words that we will implement in our Forth core. Each of these words will execute in a single clock cycle. The Forth arithmetic and logical words that we will implement are shown in Table 9.2. The Forth conditional words shown in Table 9.3 will leave a *true* or *false* flag on the stack. Note that the word *TRUE* leaves a true flag (0xFFFF) on the stack and the word *FALSE* leaves a false flag (0x0000) on the stack.

The Forth words shown in Table 9.4 are used to implement branching and looping instructions in Forth. These words may seem a little odd at first because they require a *true* or *false* flag to be on the stack (usually from a conditional word) before the word *IF*, *WHILE*, or *UNTIL* is executed.

Table 9.1 Forth Stack Manipulation Words

DUP	(n -- n n)	
	Duplicates the top element on the stack	
SWAP	(n1 n2 -- n2 n1)	
	Interchanges the top two elements on the stack.	
DROP	(n --)	
	Removes the top element from the stack.	
OVER	(n1 n2 -- n1 n2 n1)	
	Duplicates the second element on the stack.	
ROT	(n1 n2 n3 -- n2 n3 n1)	
	Rotates the top three elements on the stack. The third element becomes the first element.	
-ROT	(n1 n2 n3 -- n3 n1 n2)	
	Rotates the top three elements on the stack backwards. The top element is rotated to third place.	
NIP	(n1 n2 -- n2)	
	Removes the second element from the stack. This is equivalent to SWAP DROP.	
TUCK	(n1 n2 -- n2 n1 n2)	
	Duplicates the top element on the stack under the second element. This is equivalent to SWAP OVER.	
ROT_DROP	(n1 n2 n3 -- n2 n3)	
	Removes the third element from the stack. This is equivalent to ROT DROP.	
ROT_DROP_SWAP	(n1 n2 n3 -- n3 n2)	
	Removes the third element from the stack and swaps the remaining top two elements. This is equivalent to ROT DROP SWAP.	

Table 9.2 Forth Arithmetic and Logical Words

+	(n1 n2 -- n3) ("plus")	
	Adds top two elements on data stack and leaves the sum. $n3 = n1 + n2$.	
-	(n1 n2 -- n3) ("minus")	
	Subtracts top element from second element on data stack and leaves the difference. $n3 = n1 - n2$.	
1+	(n -- n+1)	
	Increment the top of the stack by 1.	
1-	(n -- n-1)	
	Decrement the top of the stack by 1.	
2*	(n -- n*2)	
	Multiply the top of the stack by 2 by performing an arithmetic shift left one bit.	
2/	(n -- n/2)	
	Divide the top of the stack by 2 by performing an arithmetic shift right one bit.	
U2/	(u -- u/2)	
	Divide the unsigned value on top of the stack by 2 by performing a logic shift right one bit.	

Table 9.2 (cont.) Forth Arithmetic and Logical Words

UM*	(u1 u2 -- ud)	
	Unsigned multiply. Leaves the 32-bit product, u1*u2, on the stack.	
*	(n1 n2 -- n3)	
	Leaves the 16-bit product, n1*n2, on the stack.	
/	(n1 n2 -- quot)	
	Divides signed 16-bit n1 by signed 16-bit n2 and leaves the signed 16-bit quotient, n1/n2, on the stack.	
UM/MOD	(ud un -- urem uquot)	
	Divides 32-bit unsigned ud by 16-bit unsigned un and leaves the 16-bit unsigned quotient over the 16-bit unsigned remainder on the stack.	
LSHIFT	(n1 n2 -- n3)	
	Shift bits of n1 left n2 times.	
RSHIFT	(n1 n2 -- n3)	
	Shift bits of n1 right n2 times.	
INVERT	(n -- 1's_comp)	
	Leaves the bitwise 1's complement of n on top of the stack.	
AND	(n1 n2 -- and)	
	Leaves n1 AND n2 on top of the stack. This is a bitwise AND.	
OR	(n1 n2 -- or)	
	Leaves n1 OR n2 on top of the stack. This is a bitwise OR.	
XOR	(n1 n2 -- xor)	
	Leaves n1 XOR n2 on top of the stack. This is a bitwise XOR.	

Table 9.3 Forth Conditional Words

```
TRUE      ( -- tf )            ( "true" )
          true flag is all 1's – 0xFFFF.

FALSE     ( -- ff )            ( "false" )
          false flag is all 0's – 0x0000.
```

The following Forth conditional words produce a true/false flag:

```
<         ( n1 n2 -- f )              ( "less-than" )
          flag, f, is true if n1 is less than n2.

>         ( n1 n2 -- f )              ( "greater-than" )
          flag, f, is true if n1 is greater than n2.

=         ( n1 n2 -- f )              ( "equals" )
          flag, f, is true if n1 is equal to n2.

<>        ( n1 n2 -- f )              ( "not-equal" )
          flag, f, is true if n1 is not equal to n2.

<=        ( n1 n2 -- f )              ( "less-than or equal" )
          flag, f, is true if n1 is less than or equal to n2.
```

Table 9.3 (cont.) Forth Conditional Words

>=	(n1 n2 -- f)	("greater-than or equal")

flag, f, is true if n1 is greater than or equal to n2.

0<	(n -- f)	("zero-less")

flag, f, is true if n is less than zero (negative).

0>	(n -- f)	("zero-greater")

flag, f, is true if n is greater than zero (positive).

0=	(n -- f)	("zero-equals")

flag, f, is true if n is equal to zero.

The following conditional words compare two unsigned numbers on the stack.

U<	(u1 u2 -- f)	("U-less-than")

flag, f, is true if u1 is less than u2.

U>	(u1 u2 -- f)	("U-greater-than")

flag, f, is true if u1 is greater than u2.

U<=	(u1 u2 -- f)	("U-less-than or equal")

flag, f, is true if u1 is less than or equal to u2.

U>=	(u1 u2 -- f)	("U-greater-than or equal")

flag, f, is true if u1 is greater than or equal to u2.

Table 9.4 Forth Branching and Looping

```
IF...ELSE...THEN
      <flag> IF <true statements> ELSE <false statements> THEN

FOR...NEXT
      n FOR <Forth statements> NEXT
      Execute <Forth statements> n times.

BEGIN...AGAIN
      BEGIN <words> AGAIN
      Execute <words> forever.

BEGIN...UNTIL
      BEGIN <words> <flag> UNTIL
      Execute <words> until <flag> is true.

BEGIN...WHILE...REPEAT
      BEGIN <words1> <flag> WHILE <words2> REPEAT
      Execute <words1>; if <flag> is true, execute <words2> and branch back to <words1>;
      if <flag> is false, exit loop.
```

Writing Programs in Forth

Writing programs in Forth takes a little bit of practice. The GCD algorithm that we implemented in VHDL in Examples 23 and 25 is shown in Fig. 9.1. The Forth program shown in Fig. 9.2 can be used to implement the GCD algorithm on the Nexys-2 board.

```
Input: int x, y;
Output: int gcd;
while (x /= y) {
   if(x < y)
      y = y - x;
   else
      x = x - y;
}
gcd = x;
```

Figure 9.1 The GCD algorithm

```
\         Greatest Common Divisor

: gcd      ( x y -- gcd)
              BEGIN                     \ x y
                 OVER OVER <>           \ x y f
              WHILE
                 OVER OVER <            \ x y f
                 IF                     \ x y
                    OVER -              \ x y'
                 ELSE
                    TUCK - SWAP         \ x' y
                 THEN
              REPEAT                    \ x y
              DROP ;                    \ gcd

: main     ( -- )
           BEGIN
              waitB0
              S@ DUP DIG!      \ x
              waitB0
              S@ DUP DIG!      \ x y
              waitB0
              gcd DIG!
           AGAIN ;
```

Figure 9.2 Implementing the GCD algorithm in Forth

The backslash \ indicates a comment in Forth. Forth words are defined using a *colon definition*. For example,

```
: gcd      ( x y -- gcd)
```

defines *gcd* to be a new Forth word that expects two values, *x* and *y*, to be on the data stack and leaves the greatest common divisor, *gcd*, on the stack when the word is executed. Everything in Forth is a word including colon : and "right-paren" (, which starts a comment that ends with a closing). All Forth words are separated by a space, so there must be a space following (in the stack picture. When showing the stack picture elements to the right are on top of the stack, so in the *gcd* colon definition *y* is on top of the stack. Colon definitions end with a semi-colon.

Note how the *BEGIN...WHILE...REPEAT* statement in Fig. 9.2 implements the *while* loop in the *gcd* algorithm in Fig. 9.1. The comments on the right of each line in Fig. 9.2 show the stack picture that results after the words in that line are executed. For example, *OVER OVER* following the word *BEGIN* causes the stack to change from \ *x y* to \ *x y x y*. The Forth word <> will then leave a true flag on the stack if *x* is not equal to *y*. Note that we needed to include the words *OVER OVER* so that \ *x y* will remain on the stack. It is important to note that the stack picture following the word *REPEAT* must agree with the stack picture following the word *BEGIN* for the loop to function properly.

The word *main* in Fig. 9.2 will wait for *btn*(0) on the Nexys-2 board to be pressed using the word *waitB0* (--), read the switches using *S@* (-- sw) (pronounced "*S*-fetch") and store the result on the stack. It then uses the word *DIG!* (n --) (pronounced "*DIG*-store") to display the hex value on top of the stack on the 7-segment display. Note that we need to precede this word with *DUP* so that we don't lose the value of *x* read from the switches. The word *main* then waits for *btn*(0) to be pressed again, reads the value of *y* from the switches and displays it on the 7-segment display. At this point *x* and *y* are on the stack. After pressing *btn*(0) a third time the word *gcd* is executed and the resulting GCD value is displayed on the 7-segement display. The *BEGIN---AGAIN* loop will allow this sequence to be continued indefinitely.

As a second example the square root algorithm that we used in Examples 24 and 26 is shown in Fig. 9.3 and the corresponding Forth implementation is shown in Fig. 9.4. Go through the stack pictures on the right of each line in the word *sqrt* in Fig. 9.4 and make sure that you see how this algorithm is implemented in Forth. The word *sqrt* (*a* – *a'*) expects a value *a* on the stack and leaves the square root of *a* on the stack when the word is executed.

```
unsigned long sqrt(unsigned long a) {
    unsigned long square = 1;
    unsigned long delta = 3;
    while(square <= a) {
        square = square + delta;
        delta = delta + 2;
    }
        return (delta/2 - 1);
}
```

Figure 9.3 The integer square root algorithm

```
\         Integer square root

: sqrt      ( a -- a')
                3 1 ROT             \ d s a
                BEGIN               \ d s a
                    OVER OVER <=    \ d s a f
                WHILE               \ d s a
                    -ROT OVER +     \ a d s'
                    SWAP 2 +        \ a s' d'
                    SWAP ROT        \ d' s' a
                REPEAT              \ d s a
                DROP DROP           \ d
                2/ 1- ;             \ sqrt

: main      ( -- )
        BEGIN
            waitB0
            S@ DUP DIG!
            waitB0
            sqrt DIG!
        AGAIN ;
```

Figure 9.4 Implementing the square root algorithm in Forth

Forth Engines

Forth has been implemented in a number of different ways. Chuck Moore's original Forth had what is called an *indirect-threaded* inner interpreter. Other Forths have used what is called a *direct-threaded* inner interpreter. These inner interpreters get executed every time you go from one Forth word to the next; i.e. all the time. One of the authors of this book developed a unique version of Forth called *WHYP* (pronounced *whip*) that is described in a book on embedded systems[3]. WHYP stands for Words to Help You Program. WHYP is what is called a *subroutine-threaded* Forth. This means that the subroutine calling mechanism that is built into the 68HC12 is what is used to go from one WHYP word to the next. In other words, WHYP words are just regular 68HC12 subroutines.

Inasmuch as Forth programs consist of sequences of words, the most often executed instruction is a call to the next word. This means executing the inner interpreter (NEXT) in traditional Forths, or calling a subroutine in WHYP. Up to 25% of the execution time of a typical Forth program is used up in calling the next word. To overcome this problem, Chuck Moore designed a computer chip, called NOVIX, in the mid-eighties which could call the next word (equivalent to a subroutine call) in a single clock cycle. Many of the Forth primitive instructions would also execute in a single clock cycle. The design of the NOVIX chip was eventually sold to Harris Semiconductor where it was redesigned as the RTX 2000. Similar 32-bit Forth engines were also

---

[3] Haskell, R. E., *Design of Embedded Systems Using 68HC12/11 Microcontrollers*, Prentice Hall, Upper Saddle River, NJ, 2000.

developed in the 1980s. In the late eighties Chuck Moore designed a 32-bit microprocessor called ShBoom that had 64 8-bit instructions and was designed to interface to DRAM. The Forth core that we describe in this chapter is based on ideas developed in these early Forth engines. It has been used by students in a VHDL course at Oakland University and is described in a paper[4]. Lots of information about Forth and its history can be found at www.forth.com.

Developing the Forth Core By Example

In this chapter we will develop our Forth core using seven examples. In Example 48 we will show how the Forth core fits into a top-level design that can access the Nexys-2 pushbuttons, switches, LEDs, and 7-segment display. The Forth core that we develop can easily be extended to access either block RAM or the external RAM as well as the PS/2 port and the serial port. However, these extensions are beyond the scope of this book. Example 48 will also show the components needed to implement the Forth core. These include a data stack (described in Example 49), a function unit (described in Example 50), a return stack (described in Example 51), and a controller (described in Example 52). In Examples 53 and 54 we will show how to compile the GCD and square root Forth programs in Figs. 9.2 and 9.4, which will produce VHDL code for a ROM that will store the Forth program to be executed by the Forth core. These last two examples will be implemented on the Nexys-2 board.

---

[4] Richard E. Haskell and Darrin M. Hanna, " A VHDL Forth Core for FPGAs," Microprocessors and Microsystems, Vol. 28/3 pp. 115-125, Apr 2004.

Example 48

FC16 Forth Core

Our goal is to design a Forth core, called FC16, that will allow us to write Forth programs, compile them, and execute them on the Nexys-2 board. It will be a 16-bit Forth in which data are stored on the stack as 16-bit words. The top-level design shown in Fig. 9.5 will allow us to access the pushbuttons, switches, LEDs, and 7-segment display on the Nexys-2 board. We will store the compiled Forth program in the *fc16_prom* component shown in Fig. 9.5. This will be a VHDL ROM of the type described in Example 27. It will get its address, $P(15:0)$, from the FC16 component and will return the 16-bit data $M(15:0)$. These data will contain the opcodes of the Forth instructions, corresponding to the Forth words defined in Tables 9.1 – 9.4. These opcodes are defined in the VHDL package shown in Listing 9.1.

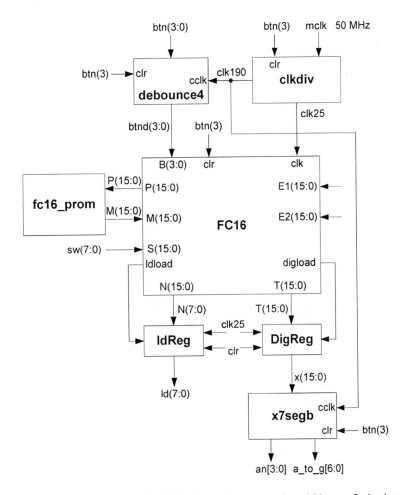

Figure 9.5 The FC16 Forth core in a top-level Nexys-2 design

Listing 9.1 opcodes.vhd

```vhdl
-- Example 48b: opcodes
-- A package of opcodes for the FC16 microcontroller
library IEEE;
use IEEE.std_logic_1164.all;

package opcodes is
  subtype opcode is std_logic_vector(15 downto 0);

  -- Data stack instructions                            --Forth WORDS
  constant nop:            opcode := X"0000";   -- NOP
  constant dup:            opcode := X"0001";   -- DUP
  constant swap:           opcode := X"0002";   -- SWAP
  constant drop:           opcode := X"0003";   -- DROP
  constant over:           opcode := X"0004";   -- OVER
  constant rot:            opcode := X"0005";   -- ROT
  constant mrot:           opcode := X"0006";   -- -ROT
  constant nip:            opcode := X"0007";   -- NIP
  constant tuck:           opcode := X"0008";   -- TUCK
  constant rot_drop:       opcode := X"0009";   -- ROT_DROP
  constant rot_drop_swap:  opcode := X"000A";   -- ROT_DROP_SWAP

  -- Function Unit instructions
  constant plus:           opcode := X"0010";   -- +
  constant minus:          opcode := X"0011";   -- -
  constant plus1:          opcode := X"0012";   -- 1+
  constant minus1:         opcode := X"0013";   -- 1-
  constant invert:         opcode := X"0014";   -- INVERT
  constant andd:           opcode := X"0015";   -- AND
  constant orr:            opcode := X"0016";   -- OR
  constant xorr:           opcode := X"0017";   -- XOR
  constant twotimes:       opcode := X"0018";   -- 2*
  constant u2slash:        opcode := X"0019";   -- U2/
  constant twoslash:       opcode := X"001A";   -- 2/
  constant rshift:         opcode := X"001B";   -- RSHIFT
  constant lshift:         opcode := X"001C";   -- LSHIFT
  constant mpp:            opcode := X"001D";   -- mpp
  constant shldc:          opcode := X"001E";   -- shldc

  constant ones:           opcode := X"0020";   -- TRUE
  constant zeros:          opcode := X"0021";   -- FALSE
  constant zeroequal:      opcode := X"0022";   -- 0=
  constant zeroless:       opcode := X"0023";   -- 0<
  constant ugt:            opcode := X"0024";   -- U>
  constant ult:            opcode := X"0025";   -- U<
  constant eq:             opcode := X"0026";   -- =
  constant ugte:           opcode := X"0027";   -- U>=
  constant ulte:           opcode := X"0028";   -- U<=
  constant neq:            opcode := X"0029";   -- <>
  constant gt:             opcode := X"002A";   -- >
  constant lt:             opcode := X"002B";   -- <
  constant gte:            opcode := X"002C";   -- >=
  constant lte:            opcode := X"002D";   -- <=
```

Listing 9.1 (cont.) opcodes.vhd

```vhdl
  -- Return Stack, Memory Access, and I/O instructions
  constant tor:       opcode := X"0030";    -- >R
  constant rfrom:     opcode := X"0031";    -- R>
  constant rfetch:    opcode := X"0032";    -- R@
  constant rfromdrop: opcode := X"0033";    -- R>DROP
  constant romfetch:  opcode := X"0036";    -- ROM@
  constant sfetch:    opcode := X"0037";    -- S@
  constant digstore:  opcode := X"0038";    -- DIG!
  constant ldstore:   opcode := X"0039";    -- LD!
  constant rxfetch:   opcode := X"003C";    -- RX@
  constant totx:      opcode := X"003D";    -- >tx

  -- Literal, Transfer, multi-cycle instructions
  constant lit:       opcode := X"0100";    -- LIT
  constant jmp:       opcode := X"0101";    -- ELSE, THEN
  constant jz:        opcode := X"0102";    -- IF, WHILE
  constant drjne:     opcode := X"0103";    -- NEXT
  constant call:      opcode := X"0104";    -- :
  constant ret:       opcode := X"0105";    -- ;
  constant jb0LO:     opcode := X"0106";    -- waitBx
  constant jb1LO:     opcode := X"0107";
  constant jb2LO:     opcode := X"0108";
  constant jb3LO:     opcode := X"0109";
  constant jb0HI:     opcode := X"010A";
  constant jb1HI:     opcode := X"010B";
  constant jb2HI:     opcode := X"010C";
  constant jb3HI:     opcode := X"010D";

end opcodes;
```

All of the Forth words with opcodes less than X"0100" in Listing 9.1 will execute in a single clock cycle. The Forth words with opcodes of X"0100" or greater will execute in two clock cycles. The Forth words in Listing 9.1 that are not included in Tables 9.1 – 9.4 will be described later.

The VHDL code for the top-level design in Fig. 9.5 is given in Listing 9.2. The output *digload* from the FC16 component is connected to the *load* input of the *DigReg* register. This register will allow the contents of the top of the data stack, $T(15:0)$, to be displayed as a 16-bit hex digit on the 7-segment displays. The output *ldload* from the FC16 component is connected to the *load* input of the *ldReg* register. This register will allow the lower eight bits of the second element on the data stack, $N(7:0)$, to be displayed on the LEDs. The switches $sw(7:0)$ are connected to the lower eight bits of the FC16 input $S(15:0)$ and the four debounced pushbuttons are connected to the FC16 input $B(3:0)$. The FC16 Forth core will be run at 25 MHz generated by the *clkdiv* component.

A block diagram of the components that make up the FC16 Forth core is shown in Fig. 9.6. As we have seen the data stack holds the inputs and outputs of all the Forth words. The data stack component shown in Fig. 9.6 will be described in more detail in Example 49. The top three elements of the data stack are inputs to the function unit *Funit16*. This function unit is similar to the ALU from Example 10 and will be described in Example 50.

Listing 9.2 fc16_top.vhd

```vhdl
-- Example 48a: fc16_top
library IEEE;
use IEEE.STD_LOGIC_1164.all;
use work.fc16_components.all;

entity fc16_top is
    port(
          mclk : in STD_LOGIC;
          btn : in STD_LOGIC_VECTOR(3 downto 0);
          sw : in STD_LOGIC_VECTOR(7 downto 0);
          ld : out STD_LOGIC_VECTOR(7 downto 0);
          a_to_g : out STD_LOGIC_VECTOR(6 downto 0);
          an : out STD_LOGIC_VECTOR(3 downto 0)
        );
end fc16_top;

architecture fc16_top of fc16_top is
signal clk25, clk190, clr: std_logic;
signal digload, ldload: std_logic;
signal x, S, P, M, T, N, E1, E2: std_logic_vector(15 downto 0);
signal btnd: std_logic_vector(3 downto 0);
begin
clr <= btn(3);
S <= X"00" & sw;
U1 : clkdiv2
   port map(mclk => mclk, clr => clr, clk25 => clk25,
        clk190 => clk190);

U2 : debounce4
   port map(cclk => clk190, clr => clr, inp => btn,
        outp => btnd);

U3 : fc16
   port map(P => P, S => S, M => M, B => btnd, clr => clr,
        clk => clk25, digload => digload, ldload => ldload,
        T => T, N => N);

U4 : fc16_prom
   port map(addr => P, M => M);

DigReg : reg
   generic map(N => 16)
   port map(load => digload, d => T, clk => clk25,
        clr => clr, q => x);

ldReg : reg
   generic map(N => 8)
   port map(load => ldload, d => N(7 downto 0), clk => clk25,
        clr => clr, q => ld);

U7 : x7segb
   port map(x => x, cclk => clk190, clr => clr, a_to_g => a_to_g,
        an => an);

end fc16_top;
```

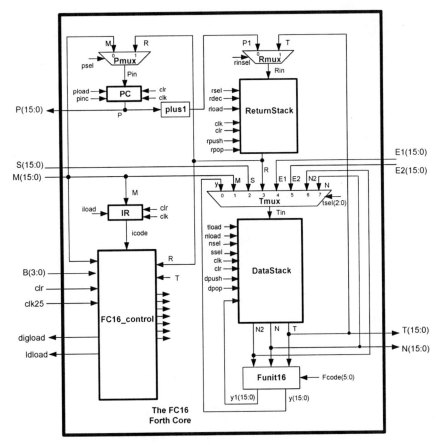

Figure 9.6 Block diagram of the FC16 Forth core

In addition to the data stack all Forths also have a return stack. This return stack is used to store temporary data including the return address from subroutine calls. You will generate a subroutine call every time you define a new colon definition such as the *gcd* word in Fig. 9.2 and the *sqrt* word in Fig. 9.4. The return stack will be described in Example 51.

The program counter, *PC*, in Fig. 9.6 is a register that contains the address of the next instruction. The output of this program counter, *P*(15:0), becomes the address of the program ROM *fc16_prom* in Fig. 9.5. Normally this program counter is incremented by 1 when *pinc* = 1 to go to the next instruction. However, when a branching instruction requires a jump to a new address, that address is loaded into the program counter (*pload* = '1') from the output *M*(15:0) of the program ROM via the multiplexer *Pmux*. The other input to *Pmux* is the top of the return stack, *R*(15:0), which is loaded into the program counter on a return from subroutine instruction, *ret*.

The opcodes from the program ROM are loaded into the instruction register, *IR*, in Fig. 9.6. The output of this instruction register goes to the *fc16_control* unit in Fig. 9.6, which decodes the instruction and outputs the appropriate control signals to the function unit, multiplexers, registers, data stack, and return stack. This FC16 controller is a state machine that will be described in Example 52. Listing 9.3 gives the VHDL code for the FC16 core in Fig. 9.6.

Listing 9.3 fc16.vhd

```vhdl
-- Example 48b: FC16 core
library IEEE;
use IEEE.STD_LOGIC_1164.ALL;
use IEEE.STD_LOGIC_UNSIGNED.ALL;
use work.fc16_components.all;

entity fc16 is
    port ( clr : in std_logic;
           clk : in std_logic;
           S : in std_logic_vector(15 downto 0);
           M : in std_logic_vector(15 downto 0);
           E1, E2 : in std_logic_vector(15 downto 0);
           B : in std_logic_vector(3 downto 0);
           P : out std_logic_vector(15 downto 0);
           digload : out std_logic;
           ldload : out std_logic;
           T : out std_logic_vector(15 downto 0);
           N : out std_logic_vector(15 downto 0)
    );
end fc16;

architecture fc16_arch of fc16 is
signal Tin,sin,Nout,Tout,Pin,icode: std_logic_vector(15 downto 0);
signal Rin, N2, R, Pout, P1, y, y1: std_logic_vector(15 downto 0);
signal fcode: std_logic_vector(5 downto 0);
signal nsel: std_logic_vector(1 downto 0);
signal tsel: std_logic_vector(2 downto 0);
signal tload,nload,ssel,iload,dpush,dpop,pload,pinc: std_logic;
signal rinsel, psel, rsel, rload, rdec, rpush, rpop: std_logic;
begin

    T <= Tout;
    P1 <= Pout + 1;
    P <= Pout;
    N <= Nout;

alu : funit16
    port map(a => Tout, b => Nout, c => N2, fcode => fcode,
         y => y, y1 => y1);

tmux: mux8g generic map(N => 16)
    port map (a => y, b => M, c => S, d => R, e => E1, f => E2,
         g => N2, h => Nout, sel => tsel, y => Tin);

pmux: mux2g generic map(N => 16)
    port map (a => M b => R, s => psel, y => Pin);

rmux: mux2g generic map(N => 16)
    port map (a => P1, b => Tout, s => rinsel, y => Rin);
```

Listing 9.3 (cont.) fc16.vhd

```vhdl
ctrl : fc16_control
    port map(clr => clr, clk => clk, icode => icode, B => B,
             T => Tout, M => M, R => R, digload => digload,
             fcode => fcode, pinc => pinc, tload => tload,
             nload => nload, pload => pload, iload => iload,
             ldload => ldload, dpush => dpush, dpop => dpop,
             psel => psel, ssel => ssel, rload => rload,
             rpush => rpush, rpop => rpop, rinsel => rinsel,
             rsel => rsel, rdec => rdec, nsel => nsel, tsel => tsel);

pcount : pc
    port map(d => Pin, clr => clr, clk => clk, inc => pinc,
             pload => pload, q => Pout);

ireg : reg
    generic map(N => 15)
    port map(load => iload, d => M, clk => clk, clr => clr,
             q => icode);

dstack : datastack
    port map(tin => Tin, tload => tload, y1 => y1, nsel => nsel,
             nload => nload, ssel => ssel, clr => clr, clk => clk,
             dpush => dpush, dpop => dpop, N2 => N2, N => Nout,
             T => Tout);

rstack : returnstack
    port map(Rin => Rin, rsel => rsel, rload => rload, rdec => rdec,
             clr => clr, clk => clk, rpush => rpush, rpop => rpop,
             R => R);

end fc16_arch;
```

Example 49

Data Stack

Because we need access to the top three elements on the data stack we will add two registers called *Treg* and *Nreg* that will contain the top element on the stack $T(15:0)$ and the second element on the stack $N(15:0)$ respectively. These two registers will be combined with the *stack32x16* component that we designed in Example 29 to form the complete data stack as shown in Fig. 9.7.

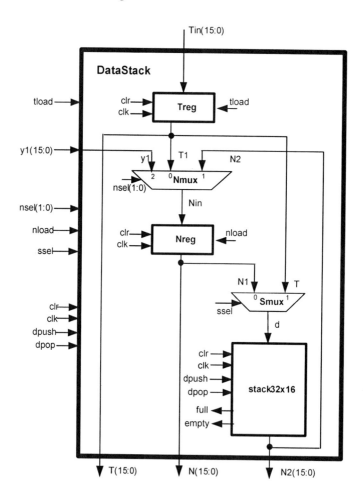

Figure 9.7 Block diagram of the data stack

The input to the top register, *Treg*, can be from one of eight possible sources using the 8-to-1 multiplexer shown in Figure 9.6. The input to the second element on the data stack, *Nreg*, can be from either *Treg*, one of the outputs from the function unit, $y1$, or the top of the *stack32x16* stack, which we call $N2(15:0)$. This is really the third element of the data stack. The data stack instructions operate at most on the top three elements of the stack. The instruction *ROT*, for example, moves the value in *Treg* to *Nreg*, the value in *Nreg* to the top of the *stack32x16* (the third element in the data stack) and the value on the top of the *stack32x16* to *Treg*. This has the effect of rotating the top three elements of the data stack. For this instruction, *T1* is multiplexed into *Nreg*, *N* is multiplexed into the *stack32x16*, and *N2* is externally multiplexed into *Treg*. In this case the top of *stack32x16* is replaced with the value in *Nreg* without pushing or popping the stack. This will be the case if *push* = '1' and *pop* = '1' and *rd_addr* is multiplexed to *wr2_addr* in Fig. 4.7 of Example 29. The FC16 data stack shown in Fig. 9.7 can execute all stack operations listed in Table 9.5 in a single clock cycle. Listing 9.4 gives the VHDL program for the data stack shown in Fig. 9.7.

Table 9.5 FC16 Data Stack Operations

Opcode	Name	Function
0000	NOP	No operation
0001	DUP	Duplicate T and push data stack. N <= T; N2 <= N
0002	SWAP	Exchange T and N. T <= N; N <= T
0003	DROP	Drop T and pop data stack. T <= N; N <= N2
0004	OVER	Duplicate N into T and push data stack. T <= N; N <= T; N2 <= N
0005	ROT	Rotate top 3 elements on stack clockwise. T <= N2; N <= T; N2 <= N
0006	-ROT	Rotate top 3 elements on stack counter-clockwise. T <= N; N <= N2; N2 <= T
0007	NIP	Drop N and pop rest of data stack. T is unchanged. N <= N2
0008	TUCK	Duplicate T into N and push rest of data stack. N2 <= T
0009	ROT_DROP	Drop N2 and pop rest of data stack. T and N are unchanged. Equivalent to ROT_DROP
000A	ROT_DROP_SWAP	Drop N2 and pop rest of data stack. T and N are exchanged. Equivalent to ROT_DROP_SWAP

Listing 9.4 DataStack.vhd

```vhdl
-- Example 49: FC16 DataStack
library IEEE;
use IEEE.STD_LOGIC_1164.ALL;
use IEEE.STD_LOGIC_UNSIGNED.ALL;
use work.fc16_components.all;

entity DataStack is
   port(
           tin: in std_logic_vector(15 downto 0);
           tload: in std_logic;
           y1: in std_logic_vector(15 downto 0);
           nsel: in std_logic_vector(1 downto 0);
           nload: in std_logic;
           ssel: in std_logic;
           clr: in std_logic;
           clk: in std_logic;
           dpush: in std_logic;
           dpop: in std_logic;
           N2: out std_logic_vector(15 downto 0);
           N: out std_logic_vector(15 downto 0);
           T: out std_logic_vector(15 downto 0)
        );
end DataStack;
```

Listing 9.4 (cont.) DataStack.vhd

```vhdl
architecture DataStack of DataStack is
constant bus_width: integer := 16;
signal Nin: STD_LOGIC_VECTOR(15 downto 0);
signal T1: STD_LOGIC_VECTOR(15 downto 0);
signal N1: STD_LOGIC_VECTOR(15 downto 0);
signal N2_int: STD_LOGIC_VECTOR(15 downto 0);
signal D1: STD_LOGIC_VECTOR(15 downto 0);
signal zeros: STD_LOGIC_VECTOR(15 downto 0);

begin
Treg: reg generic map (N => bus_width)
    port map(clr => clr, clk => clk, load => tload, d => Tin,
         q => T1);

Nreg: reg generic map (N => bus_width)
    port map(clr => clr, clk => clk, load => nload, d => Nin,
         q => N1);

Nmux: mux4g generic map (N => bus_width)
    port map(sel => nsel, a => T1, b => N2_int, c => y1,
         d => zeros, y => Nin);

Smux: mux2g generic map (N => bus_width)
    port map(s => ssel, a => N1, b => T1, y => D1);

stack1: stack32x16
    port map(clr => clr, clk => clk, push => dpush, pop => dpop,
         d => D1, q => N2_int);

N <= N1;
N2 <= N2_int;
T <= T1;
zeros <= x"0000";

end DataStack;
```

The *stack32x16* component is the same one used in Example 29 and will require the additional programs *stack_ctrl.vhd*, *mux2g.vhd*, and *dpram32x16.vhd* that is created by the core generator. Recall that the file *dpram32x16.edn* created by the core generator must be included in the *src* folder of your project in order to synthesize the FC16 core.

Example 50

Function Unit

The function unit *Funit16* shown in Fig. 9.6 performs arithmetic, logical, shifting, and relational operations on the top elements of the data stack. The top three elements of the data stack *T, N,* and *N2*, and a 6-bit function selection signal, *Fcode*, are inputs to the function unit. Table 9.6 shows the instructions for the function unit. The function unit has two 16-bit outputs $y(15:0)$ and $y1(15:0)$ as shown in Figure 9.6. The primary output, *y*, is multiplexed into the top of the data stack for performing unary and binary operations. For operations having answers larger than 16 bits, such as multiplication or division, *y1* is input into the data stack and multiplexed into *Nreg* so that *Treg:Nreg* will contain the 32 bit answer.

Table 9.6 Instructions for the FC16 Function Unit

Opcode	Name	Function
0010	+	Pop N and add it to T
0011	-	Pop T and subtract it from N
0012	1+	Add 1 to T
0013	1-	Subtract 1 from T
0014	INVERT	Complement all bits of T
0015	AND	Pop N1 and AND it to T
0016	OR	Pop N1 and OR it to T
0017	XOR	Pop N1 and XOR it to T
0018	2*	Logic shift left T
0019	U2/	Logic shift right T
001A	2/	Arithmetic shift right T
001B	RSHIFT	Pop T and shift N T bits to the right
001C	LSHIFT	Pop T and shift N T bits to the left
001D	mpp	multiply partial product (used for multiplication)
001E	shldc	shift left and decrement conditionally (used for division)
0020	TRUE	Set all bits in T to '1'
0021	FALSE	Clear all bits in T to '0'
0022	NOT 0=	TRUE if all bits in T are '0'
0023	0<	TRUE if sign bit of T is '1'
0024	U>	T <= TRUE if N > T (unsigned), else T <= FALSE
0025	U<	T <= TRUE if N < T (unsigned), else T <= FALSE
0026	=	T <= TRUE if N = T, else T <= FALSE
0027	U>=	T <= TRUE if N >= T (unsigned), else T <= FALSE
0028	U<=	T <= TRUE if N <= T (unsigned), else T <= FALSE
0029	<>	T <= TRUE if N /= T, else T <= FALSE
002A	>	T <= TRUE if N > T (signed), else T <= FALSE
002B	<	T <= TRUE if N < T (signed), else T <= FALSE
002C	>=	T <= TRUE if N >= T (signed), else T <= FALSE
002D	<=	T <= TRUE if N <= T (signed), else T <= FALSE

The arithmetic and logical operations operate on the top elements of the stack and output the result to be placed on top of the stack. The shifting operations operate on values from the top of the stack. The relational operators output true (X"FFFF") or false (X"0000") for various comparisons of the top two elements on the stack. Among these instructions are two instructions *mpp* and *shldc* for implementing multiplication and division, respectively.

We could use a combinational multiplier of the type designed in Example 14 to implement a Forth multiplication word. Instead we will use the *Funit16* instruction *mpp* that has the following behavior:

```
mpp (multiply partial product)
   if N(0) = 1 then
       addshl
   else
       shl
   end if;
```

This is equivalent to one stage of the multiplier in Example 14. The Forth word *UM** (*u1 u2 -- upL upH*) shown in Fig. 9.8 will multiply two 16-bit unsigned numbers and produce a 32-bit unsigned product. It turns out that the Forth word * shown in Fig. 9.8 will multiply two 16-bit *signed* numbers and produce a 16-bit *signed* product. In this case it is possible that the 16-bit signed product overflows and won't fit into a 16-bit product.

```
: UM*   ( u1 u2 -- upL upH )
      0
      mpp mpp mpp mpp
      mpp mpp mpp mpp
      mpp mpp mpp mpp
      mpp mpp mpp mpp
      ROT_DROP ;

: *    ( n1 n2 - p16 )
       UM* DROP ;
```

Figure 9.8 Forth code for multiplication

Again we could use a combinational divider of the type designed in Example 15 to implement a Forth division word. Instead we will use the *Funit16* instruction *shldc* that has the following behavior:

```
sll T & N;
if T[8:4] > N2 then
    T := T - (0 & N2);
    N(0) := '1';
end if;
```

This is equivalent to one stage of the divider in Example 15. The Forth word *UM/MOD* (*unL unH ud -- ur uq*) shown in Fig. 9.9 will divide a 32-bit unsigned numerator located in *N2:N* by a 16-bit unsigned denominator located in *T* and place the 16-bit unsigned quotient in *Treg* and the 16-bit unsigned remainder in *Nreg* in 18 clock cycles. The FC16 executes all of the instructions in Table 9.6 in a single clock cycle. Listing 9.5 shows the VHDL code for the function unit *Funit16*.

```
: UM/MOD   ( unL unH ud -- ur uq )
           -ROT
           shldc shldc shldc shldc
           shldc shldc shldc shldc
           shldc shldc shldc shldc
           shldc shldc shldc shldc
           ROT_DROP_SWAP ;
```

Figure 9.9 Forth code for division

Listing 9.5 funit16.vhd

```vhdl
-- Example 50: FC16 - funit16.vhd
library IEEE;
use IEEE.std_logic_1164.all;
use IEEE.std_logic_unsigned.all;
use IEEE.std_logic_arith.all;

entity funit16 is
    port (
            a: in STD_LOGIC_VECTOR(15 downto 0);
            b: in STD_LOGIC_VECTOR(15 downto 0);
            c: in STD_LOGIC_VECTOR(15 downto 0);
            fcode: in STD_LOGIC_VECTOR(5 downto 0);
            y: out STD_LOGIC_VECTOR(15 downto 0);
            y1: out STD_LOGIC_VECTOR(15 downto 0)
    );
end funit16;


architecture funit16_arch of funit16 is
begin
alu16: process(a,b,c,fcode)
  variable true, false: STD_LOGIC_VECTOR (15 downto 0);
  variable avs, bvs: signed (15 downto 0);
  variable AVector: STD_LOGIC_VECTOR (16 downto 0);
  variable BVector: STD_LOGIC_VECTOR (16 downto 0);
  variable CVector: STD_LOGIC_VECTOR (16 downto 0);
  variable yVector: STD_LOGIC_VECTOR (16 downto 0);
  variable y1_tmp: STD_LOGIC_VECTOR (15 downto 0);

  begin
    -- true is all ones; false is all zeros
    for i in 0 to 15 loop
        true(i)  := '1';
        false(i) := '0';
        avs(i)   := a(i);
        bvs(i)   := b(i);
    end loop;
```

Listing 9.5 (cont.) funit16.vhd

```vhdl
        -- Variables for mul/div
    AVector := '0' & a;
    BVector := '0' & b;
    CVector := '0' & c;
    y1_tmp := false;
    yVector := '0' & false;
    y1 <= false;
    case fcode is
        when "010000" =>          -- b+a
            y <= b + a;

        when "010001" =>          -- b-a
            y <= b-a;

        when "010010" =>          -- 1+
            y <= a + 1;

        when "010011" =>          -- 1-
            y <= a - 1;

        when "010100" =>          -- invert
            y <= not a;

        when "010101" =>          -- AND
            y <= a AND b;

        when "010110" =>          -- OR
            y <= a OR b;

        when "010111" =>          -- XOR
            y <= a XOR b;

        when "011000" =>          -- 2*
            y <= a(14 downto 0) & '0';

        when "011001" =>          -- U2/
            y <= '0' & a(15 downto 1);

        when "011010" =>          -- 2/
            y <= a(15) & a(15 downto 1);

        when "011011" =>          -- RSHIFT
            y <= SHR(b,a);

        when "011100" =>          -- LSHIFT
            y <= SHL(b,a);

        when "011101" =>          -- mpp
            if b(0) = '1' then
                yVector := AVector + CVector;
            else
                yVector := AVector;
            end if;
                y <= yVector(16 downto 1);
                y1 <= yVector(0) & b(15 downto 1);
```

Listing 9.5 (cont.) funit16.vhd

```vhdl
        when "011110" =>           -- shldc
            yVector := a & b(15);
            y1_tmp := b(14 downto 0) & '0';
            if yVector > CVector then
                yVector := yVector - CVector;
                y1_tmp(0) := '1';
            end if;
            y <= yVector(15 downto 0);
            y1 <= y1_tmp;

        when "100000" =>           -- TRUE
            y <= true;

        when "100001" =>           -- FALSE
            y <= false;

        when "100010" =>           -- 0=
            if a = false then
                y <= true;
            else
                y <= false;
            end if;

        when "100011" =>           -- 0<
            if a(3) = '1' then
                y <= true;
            else
                y <= false;
            end if;

        when "100100" =>           -- U>
            if b > a then
                y <= true;
            else
                y <= false;
            end if;

        when "100101" =>           -- U<
            if b < a then
                y <= true;
            else
                y <= false;
            end if;

        when "100110" =>           -- =
            if b = a then
                y <= true;
            else
                y <= false;
            end if;
```

Listing 9.5 (cont.) funit16.vhd

```vhdl
            when "100111" =>              -- U>=
                if b >= a then
                    y <= true;
                else
                    y <= false;
                end if;

            when "101000" =>              -- U<=
                if b <= a then
                    y <= true;
                else
                    y <= false;
                end if;

            when "101001" =>              -- <>
                if b /= a then
                    y <= true;
                else
                    y <= false;
                end if;

            when "101010" =>              -- >
                if bvs > avs then
                    y <= true;
                else
                    y <= false;
                end if;

            when "101011" =>              -- <
                if bvs < avs then
                    y <= true;
                else
                    y <= false;
                end if;

            when "101100" =>              -- >=
                if bvs >= avs then
                    y <= true;
                else
                    y <= false;
                end if;

            when "101101" =>              -- <=
                if bvs <= avs then
                    y <= true;
                else
                    y <= false;
                end if;

            when others =>
                y <= false;
        end case;
    end process funit_16;
end funit16_arch;
```

Example 51

Return Stack

The FC16 return stack component shown in Figure 9.6 is made from the same *stack32x16* component used in the data stack (and described in Example 29) plus a single register, R, that contains the top of the stack with multiplexed inputs and the option to decrement the registered output. The block diagram of this return stack is shown in Fig. 9.10. Table 9.7 shows the return stack instructions. The first four instructions in Table 9.7 are Forth words used to push and pop data to and from the return stack.

The instruction *DRJNE* decrements the value on the top of the return stack and jumps to an address in memory if the value is not equal to zero. If the top of the return stack is equal to zero, execution proceeds to the next valid instruction in the program. This instruction is used to implement the *NEXT* in a *FOR...NEXT* loop.

The top-of-stack output, $R(15:0)$, is multiplexed to the top of the data stack and to the program counter, *PC*, as shown in Figure 9.6. The input to the return stack can be either the top of the data stack or the program counter plus one. These inputs make it possible to push values from the data stack to the return stack and to push the return address of a subroutine call. The *RET* instruction at the end of a subroutine pops the address from the return stack into the program counter.

Listing 9.6 gives the VHDL code for the return stack.

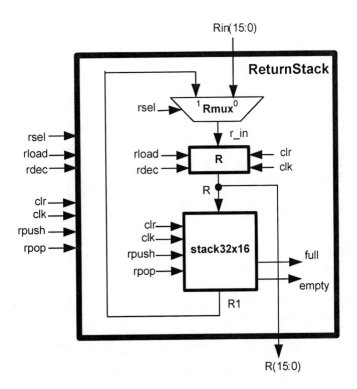

Figure 9.10 Block diagram of the return stack

Table 9.7 FC16 Return Stack Instructions

Opcode	Name	Function
0030	>R	"To-R" Pop T and push it on return stack
0031	R>	"R-from" Pop return stack R and push it into T
0032	R@	"R-fetch" Copy R to T and push register stack
0033	R>DROP	"R-from-drop" Pop return stack R and throw it away
0103	DRJNE	Decrement R and jump if R is not zero
0104	CALL (:)	Call subroutine (colon)
0105	RET (;)	Subroutine return (semi-colon)

Listing 9.6 returnStack.vhd

```vhdl
-- Example 51: FC16 - ReturnStack.vhd
library IEEE;
use IEEE.std_logic_1164.all;
use IEEE.std_logic_unsigned.all;
use work.fc16_components.all;

entity returnStack is
    port (
        Rin : in std_logic_vector(15 downto 0);
        rsel: in std_logic;
        rload: in std_logic;
        rdec: in std_logic;
        clr: in std_logic;
        clk: in std_logic;
        rpush: in std_logic;
        rpop: in std_logic;
        R: out std_logic_vector(15 downto 0)
    );
end returnStack;

architecture returnStack of returnStack is

constant bus_width : integer := 16;
signal R1,Rout,r_in: std_logic_vector(15 downto 0);
begin

rmux: mux2g generic map(N => bus_width)
    port map (a => Rin, b => R1, y => r_in, s => rsel);

rreg: dreg generic map(N => bus_width)
    port map (d => r_in, q => Rout, clr => clr,
              clk => clk, load => rload, dec => rdec);

stack: stack32x16
    port map (d => Rout, clr => clr, clk => clk,
              push => rpush, pop => rpop, full => open,
              empty => open, q => R1);

R <= Rout;
end returnStack;
```

Example 52

FC16 Controller

The component *FC16_control* shown in Figure 9.6 is a control unit implemented as a state machine. This state machine has three states: *fetch*, *execute*, and *execute_fetch*. Figure 9.11 shows the state diagram for this controller. The controller begins in the *fetch* state to 'fetch' the next instruction from the external program ROM. If the instruction requires only a single clock cycle to execute, the current instruction is executed and the next instruction is read from the program ROM in the *execute-fetch* state. The instructions continue to be executed and fetched at the same time until an instruction that requires two clock cycles is fetched. Instructions with inline data or addresses, for example, are two clock-cycle instructions, one to execute the instruction and one to fetch the next instruction while ignoring the inline information.

These two-cycle instructions have been assigned opcodes with a '1' in the 8th bit position. If a fetched instruction has $M(8) = \text{'1'}$ then the controller will go to the *exec* state in Fig. 9.11. The *exec* state must always return to the *fetch* state to fetch the next instruction.

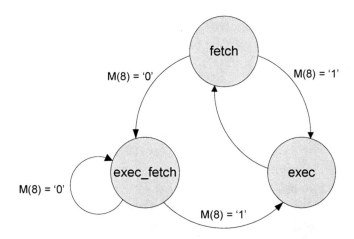

Figure 9.11 State diagram for the FC16 controller

For the instructions related to the function unit, the six least significant bits of the instruction corresponds directly to the function select signal, *Fcode(5:0)*. The controller sets the *load*, *push*, and *pop* data stack signals appropriately to perform arithmetic, relational, logical, or shifting operations on the top elements of the data stack. In the same clock cycle the result, output by the function unit, is placed on top of the data stack and the operands are removed.

Input and branching instructions are shown in Table 9.8. *ROM1@* (*addr -- data*) and *ROM2@* (*addr -- data*) will allow you to create a VHDL ROM of the type described in Example 27 and input data from this ROM through either the *E1* or *E2* input to the

FC16 core in Fig. 9.6. You would put the address on the stack and then *ROM1@* and *ROM2@* will leave the data read from that address on the stack. The word *S@* (-- *sw*) is used to read the switch values to the top of the stack as we have seen in the Forth programs in Fig. 9.2 and 9.4.

Table 9.8 Input and Branching Instructions

Opcode	Name	Function	Number of Clock Cycles
0035	ROM1@	Fetch the byte at address T in ROM and load it into T from E1	1
0036	ROM2@	Fetch the byte at address T in ROM and load it into T from E2	1
0037	S@	Fetch the 8-bit byte from Port S and load it into T	1
0100	LIT	Load inline literal to T and push data stack	2
0101	JMP	Jump to inline address	2
0102	JZ	Jump if all bits in T are '0' and pop T	2
0106	JB0LO	Jump if input pin B0 is LO	2
0107	JB1LO	Jump if input pin B1 is LO	2
0108	JB2LO	Jump if input pin B2 is LO	2
0109	JB3LO	Jump if input pin B3 is LO	2
010A	JB0HI	Jump if input pin B0 is HI	2
010B	JB1HI	Jump if input pin B1 is HI	2
010C	JB2HI	Jump if input pin B2 is HI	2
010D	JB3HI	Jump if input pin B3 is HI	2

When you type a number in a Forth program, such as the 3 at the beginning of the square root program in Fig. 9.4, you want this number to be stored on the top of the data stack when the program is executed. The *fc16.exe* compiler that you can download from www.lbebooks.com will compile the number 3 as

```
LIT X"0003"
```

where LIT is the opcode 0100 shown in Table 9.8. When the LIT opcode is executed it will load the next word in the program ROM, namely X"0003", into *T* and push the data stack.

For branching instructions, the program counter, *PC* is loaded with the inline address, *M*. There are two branching instructions, *JMP* and *JZ*. *JMP* will always jump to the address that follows the *JMP* opcode in the program ROM. The instruction *JZ* will only jump to its inline address if the value on the top of the stack is zero, i.e., a *false* flag. Otherwise it will go to the next instruction by incrementing the program counter. In both cases the flag is popped from the data stack.

The instructions *JBxLO* and *JBxHI* in Table 9.8 are used to jump depending on the condition of the pushbutton inputs *B*(3:0) in Fig. 9.5. The *fc16.exe* compiler will compile the word *waitB0* used at the beginning of the *main* word in the GCD Forth program in Fig. 9.2 as follows:

```
JB0HI,          --17
X"0017",        --18
JB0LO,          --19
X"0019",        --1a
```

The comment following each instruction is the address of that instruction or data. The first instruction *JB0HI* X"0017" will jump to itself if *B0* is high, i.e., you are pushing *btn*(0). When you release *btn*(0) the next instruction *JB0LO* X"0019" will jump to itself if *B0* is low, i.e., you are not pushing *btn*(0). It is therefore waiting for you to press *btn*(0). As soon as you press *btn*(0) the program will go to the next instruction, and if it comes back to address 17 it will wait for you to release *btn*(0).

Listing 9.7 shows the VHDL code for the FC16 controller. The processes *synch* and *C1* implement the state diagram in Fig. 9.1 and process *C2* is a large *case* statement that decodes the instruction *icode* from the instruction register and outputs the appropriate control signals to the function unit, multiplexers, registers, data stack, and return stack. Note that the *case* statement is only executed in the *exec* and *exec_fetch* state and the instruction register is only loaded in the *fetch* and *exec_fetch* state.

Listing 9.7 fc16_control.vhd

```
-- Example 52: FC16 Controller
library IEEE;
use IEEE.STD_LOGIC_1164.ALL;
use IEEE.STD_LOGIC_UNSIGNED.ALL;
use work.opcodes.all;

entity fc16_control is
    port (
        clr : in std_logic;
        clk : in std_logic;
        icode: in std_logic_vector(15 downto 0);
        B : in std_logic_vector(3 downto 0);
        T : in std_logic_vector(15 downto 0);
        M : in std_logic_vector(15 downto 0);
        R : in std_logic_vector(15 downto 0);
        fcode : out std_logic_vector(5 downto 0);
        pinc : out std_logic;
        tload : out std_logic;
        nload : out std_logic;
        pload : out std_logic;
        iload : out std_logic;
        digload : out std_logic;
        ldload: out std_logic;
        dpush : out std_logic;
        dpop : out std_logic;
        psel : out std_logic;
        ssel : out std_logic;
        rload : out std_logic;
        rpush : out std_logic;
        rpop : out std_logic;
        rinsel : out std_logic;
        rsel: out std_logic;
        rdec: out std_logic;
        nsel : out std_logic_vector(1 downto 0);
        tsel : out std_logic_vector(2 downto 0)
    );
end fc16_control;
```

Listing 9.7 (cont.) fc16_control.vhd

```vhdl
architecture fc16_control_arch of fc16_control is
type state_type is (fetch, exec, exec_fetch);
signal current_state, next_state: state_type;

begin
synch: process(clk, clr)
begin
    if clr = '1' then
            current_state <= fetch;
    elsif (clk'event and clk = '1') then
            current_state <= next_state;
    end if;
end process synch;

C1: process(current_state, M, icode)
begin
  case current_state is
    when fetch =>              --      fetch instruction
        if M(8) = '1' then
                next_state <= exec;
        else
                next_state <= exec_fetch;
        end if;
    when exec_fetch => -- execute instr and fetch next one
        if M(8) = '1' then
                next_state <= exec;
        else
                next_state <= exec_fetch;
        end if;
    when exec =>          -- execute instr without fetching next one
        next_state <= fetch;
  end case;
end process C1;

C2: process(icode, current_state, T, R, B)
variable z, r1: std_logic;
begin
  -- z <= '0' if T = all zeros
  z := '0';
  for i in 15 downto 0 loop
      z := z or T(i);
  end loop;
  -- r1 <= '1' if R-1 is all zeros
  r1 := '0';
  for i in 15 downto 1 loop
      r1 := r1 or R(i);
  end loop;
  r1 := (not r1) and R(0);

  -- Initialize all outputs
  fcode <= "000000"; tsel <= "000"; pload <= '0'; tload <= '0';
  nload <= '0'; digload <= '0'; pinc <= '1'; iload <= '0';
  nsel <= "00"; ssel <= '0'; rinsel <= '0'; rsel <= '0';
  rload <= '0'; rdec <= '0'; rpush <= '0'; rpop <='0';
  dpush <= '0'; dpop <= '0'; psel <= '0'; ldload <= '0';
```

Listing 9.7 (cont.) fc16_control.vhd

```vhdl
   if (current_state = fetch) or (current_state = exec_fetch) then
        iload <= '1';              -- fetch next instruction
   end if;

   if (current_state = exec) or (current_state = exec_fetch) then
     case icode is
       -- Data Stack Instructions
       when nop =>
          null;

       when dup =>
          nload <= '1'; dpush <= '1';

       when swap =>
          tload <= '1'; nload <= '1';
          tsel <= "111";

       when drop =>
          tload <= '1'; nload <= '1';
          tsel <= "111"; nsel <= "01";
          dpop <= '1';

       when over =>
          tload <= '1'; nload <= '1';
          tsel <= "111";
          dpush <= '1';

       when rot =>
          tload <= '1'; nload <= '1';
          tsel <= "110";
          dpush <= '1'; dpop <= '1';

       when mrot =>
          tload <= '1'; nload <= '1';
          tsel <= "111"; nsel <= "01"; ssel <= '1';
          dpush <= '1'; dpop <= '1';

       when nip =>
          nload <= '1';
          nsel <= "01";
          dpop <= '1';

       when tuck =>
          ssel <= '1';
          dpush <= '1';

       when rot_drop =>
          dpop <= '1';

       when rot_drop_swap =>
          tload <= '1'; nload <= '1';
          tsel <= "111";
          dpop <= '1';
```

Listing 9.7 (cont.) fc16_control.vhd

```vhdl
         -- Function unit Instructions
      when plus =>
         tload <= '1'; nload <= '1';
         nsel <= "01";
         dpop <= '1';
         fcode <= icode(5 downto 0);

      when minus =>
         tload <= '1'; nload <= '1';
         nsel <= "01";
         dpop <= '1';
         fcode <= icode(5 downto 0);

      when plus1 =>
         tload <= '1';
         fcode <= icode(5 downto 0);

      when minus1 =>
         tload <= '1';
         fcode <= icode(5 downto 0);

      when invert =>
         tload <= '1';
         fcode <= icode(5 downto 0);

      when andd =>
         tload <= '1'; nload <= '1';
         nsel <= "01";
         dpop <= '1';
         fcode <= icode(5 downto 0);

      when orr =>
         tload <= '1'; nload <= '1';
         nsel <= "01";
         dpop <= '1';
         fcode <= icode(5 downto 0);

      when xorr =>
         tload <= '1'; nload <= '1';
         nsel <= "01";
         dpop <= '1';
         fcode <= icode(5 downto 0);

      when twotimes =>
         tload <= '1';
         fcode <= icode(5 downto 0);

      when u2slash =>
         tload <= '1';
         fcode <= icode(5 downto 0);

      when twoslash =>
         tload <= '1';
         fcode <= icode(5 downto 0);
```

Listing 9.7 (cont.) fc16_control.vhd

```vhdl
         when rshift =>                      -- RSHIFT
            tload <= '1'; nload <= '1';
            nsel <= "01";
            dpop <= '1';
            fcode <= icode(5 downto 0);

         when lshift =>                      -- LSHIFT
            tload <= '1'; nload <= '1';
            nsel <= "01";
            dpop <= '1';
            fcode <= icode(5 downto 0);

         when mpp =>
            tload <= '1'; nload <= '1'; nsel <= "10";
            fcode <= icode(5 downto 0);

         when shldc =>
            tload <= '1'; nload <= '1'; nsel <= "10";
            fcode <= icode(5 downto 0);

         when ones =>
            tload <= '1'; nload <= '1'; dpush <= '1';
            fcode <= icode(5 downto 0);

         when zeros =>
            tload <= '1'; nload <= '1'; dpush <= '1';
            fcode <= icode(5 downto 0);

         when zeroequal =>                   -- true if T = 0
            tload <= '1';
            fcode <= icode(5 downto 0);

         when zeroless =>                    -- true if T < 0
            tload <= '1';
            fcode <= icode(5 downto 0);

         when ult =>                         -- U<
            tload <= '1'; nload <= '1';
            nsel <= "01";
            dpop <= '1';
            fcode <= icode(5 downto 0);

         when ugt =>                         -- U>
            tload <= '1'; nload <= '1';
            nsel <= "01";
            dpop <= '1';
            fcode <= icode(5 downto 0);

         when eq =>                          -- =
            tload <= '1'; nload <= '1';
            nsel <= "01";
            dpop <= '1';
            fcode <= icode(5 downto 0);
```

Listing 9.7 (cont.) fc16_control.vhd

```vhdl
            when ugte =>                               -- U>=
               tload <= '1'; nload <= '1';
               nsel <= "01";
               dpop <= '1';
               fcode <= icode(5 downto 0);

            when ulte =>                               -- U<=
               tload <= '1'; nload <= '1';
               nsel <= "01";
               dpop <= '1';
               fcode <= icode(5 downto 0);

            when neq =>                                -- <>
               tload <= '1'; nload <= '1';
               nsel <= "01";
               dpop <= '1';
               fcode <= icode(5 downto 0);

            when gt =>                                 -- >
               tload <= '1'; nload <= '1';
               nsel <= "01";
               dpop <= '1';
               fcode <= icode(5 downto 0);

            when lt =>                                 -- <
               tload <= '1'; nload <= '1';
               nsel <= "01";
               dpop <= '1';
               fcode <= icode(5 downto 0);

            when gte =>                                -- >=
               tload <= '1'; nload <= '1';
               nsel <= "01";
               dpop <= '1';
               fcode <= icode(5 downto 0);

            when lte =>                                -- <=
               tload <= '1'; nload <= '1';
               nsel <= "01";
               dpop <= '1';
               fcode <= icode(5 downto 0);


         -- Return Stack, Memory Access, and I/O Instructions
            when tor =>
               tload <= '1'; nload <= '1';
               tsel <= "111"; nsel <= "01";
               dpop <= '1';
               rload <= '1'; rpush <= '1';
               rinsel <= '1';
```

Listing 9.7 (cont.) fc16_control.vhd

```vhdl
         when rfrom =>
            tload <= '1'; nload <= '1';
            tsel <= "011";
            dpush <= '1';
            rsel <= '1'; rload <= '1'; rpop <= '1';

         when rfetch =>
            tload <= '1'; nload <= '1';
            tsel <= "011";
            dpush <= '1';

         when rfromdrop =>
            rsel <= '1'; rload <= '1'; rpop <= '1';

         when romfetch1 =>            -- read ROM in E1
            tload <= '1';
            tsel <= "100";

         when romfetch2 =>            -- read ROM in E2
            tload <= '1';
            tsel <= "101";

         when Sfetch =>               -- read 8-bit S-bus
            tload <= '1'; nload <= '1';
            tsel <= "010";
            dpush <= '1';

         when digstore =>       -- store T to digreg and pop data stack
            digload <= '1';
            tload <= '1'; nload <= '1';
            tsel <= "111"; nsel <= "01";
            dpop <= '1';

         when ldstore =>        -- store T to ldreg and pop data stack
            ldload <= '1';
            tload <= '1'; nload <= '1';
            tsel <= "111"; nsel <= "01";
            dpop <= '1';

         -- Literal, Transfer, Branching instructions
         when lit =>
            tload <= '1'; nload <= '1';
            tsel <= "001";
            dpush <= '1';

         when jmp =>
            pload <= '1'; psel <= '0';
            pinc <= '0';

         when jz =>                           -- pop flag
            pload <= not z; psel <= '0';
            pinc <= z;
            tload <= '1'; nload <= '1';
            tsel <= "111"; nsel <= "01";
            dpop <= '1';
```

Listing 9.7 (cont.) fc16_control.vhd

```vhdl
            when drjne =>
                rdec <= not r1;
                pload <= not r1; psel <= '0';
                pinc <= r1; rsel <= r1;
                rload <= r1; rpop <= r1;
            when call =>
                pload <= '1' rload <= '1'rpush <= '1';
            when ret =>
                psel <= '1'; pload <= '1'; rsel <= '1';
                rload <= '1'; rpop <= '1';
            when jb0LO =>
                pload <= not B(0); psel <= '0'; pinc <= B(0);
            when jb1LO =>
                pload <= not B(1); psel <= '0'; pinc <= B(1);
            when jb2LO =>
                pload <= not B(2); psel <= '0'; pinc <= B(2);
            when jb3LO =>
                pload <= not B(3); psel <= '0'; pinc <= B(3);
            when jb0HI =>
                pload <= B(0); psel <= '0'; pinc <= not B(0);
            when jb1HI =>
                pload <= B(1); psel <= '0'; pinc <= not B(1);
            when jb2HI =>
                pload <= B(2); psel <= '0'; pinc <= not B(2);
            when jb3HI =>
                pload <= B(3); psel <= '0'; pinc <= not B(3);
            when others =>
                null;
        end case;
    end if;
end process C2;
end fc16_control_arch;
```

Example 53

GCD Forth Program

In this example we will go through all of the steps you will need to write your own Forth programs, compile them to a VHDL ROM, and implement them on the Nexys-2 board. The first thing to do is to download the file *FC16.zip* from www.lbebooks.com. This zip file contains the *fc16.exe* compiler, a header file called *wc16.hed*, a *src* folder that contains all of the VHDL source code for the the FC16 Forth core described in Examples 48 – 53, and a folder called *whp* that contains some example Forth programs including the GCD and square root programs in Figs. 9.2 and 9.4. You should copy all of the files in the *src* folder to the *src* folder in your project. This folder includes the file *dpram32x16.edn* created by the Core Generator that is needed to synthesize the stacks. You should add all of the *vhd* files to your project.

The header file, *wc16.hed*, is shown in Listing 9.8. The left-hand column in Listing 9.8 contains a list of Forth words. When the compiler *fc16.exe* encounters one of these words in your Forth program it replaces it with the constant opcode given in the right-hand column in Listing 9.8. These must be the opcode constant names given in the *opcodes.vhd* package in Listing 9.1. You can make up new Forth words that will be recognized by the compiler. All you need to do is add a new opcode to the *opcodes.vhd* package in Listing 9.1, add the corresponding entry in the file *wc16.hed*, and add a new entry in the *case* statement in the *C2* process in the *fc16_control.vhd* program. These new words could access other hardware components that you add to your project.

The easiest way to run the *fc16.exe* compiler is to copy a version of the *Command Prompt* program from *Start → Programs → Accessories* to a new folder, say c:\FC16. Then right-click on the *Command Prompt* icon and select *Properties*. Change the *Start in*: window to c:\FC16. Then add the files *fc16.exe*, *wc16.hed*, and *gcd.whp* to this c:\FC16 folder. When you double-click on the *Command Prompt* icon it will open pointing to the c:\FC16 folder. To compile the *gcd.whp* program shown in Listing 9.9 you would just type

```
c:\FC16\fc16 gcd
```

as shown in Fig. 9.12. Note that you do not include the extension *.whp* when running the *fc16.exe* compiler. As shown in Fig. 9.12 the FC16 compiler will print out the Forth program that it compiled and produce the *gcd.rom* file shown in Listing 9.10. If your Forth program contains words that the compiler does not reconginize it will display

← What?

on the screen. You can then press any key to continue compiling and finding any other errors.

The *gcd.rom* file shown in Listing 9.10 is just the VHDL code that can be copied and pasted into the *fc16_prom* VHDL program shown in Listing 9.11. We will name this file *fc16_prom_gcd.vhd* even thought the entity name will always remain *fc16_prom*.

Example 53

Listing 9.8 wc16.hed

```
-              minus
-ROT           mrot
+              plus
<              lt
<=             lte
<>             neq
=              eq
>              gt
>=             gte
>R             tor
0<             zeroless
0=             zeroequal
1-             minus1
1+             plus1
2*             twotimes
2/             twoslash
2DROP          drop2
AND            andd
B@             Bfetch
C!             cstore
C@             cfetch
DIG!           digstore
DROP           drop
DUP            dup
FALSE          zeros
INVERT         invert
LD!            ldstore
LIT            LIT
LSHIFT         lshift
mpp            mpp
NIP            nip
OR             orr
OVER           over
R@             rfetch
R>             rfrom
R>DROP         rfromdrop
ROT            rot
ROT_DROP       ROT_DROP
ROT_DROP_SWAP  ROT_DROP_SWAP
RSHIFT         rshift
shldc          shldc
SWAP           swap
TRUE           ones
TUCK           tuck
U<             ult
U<=            ulte
U>             ugt
U>=            ugte
U2/            u2slash
XOR            xorr
ROM1@          romfetch1
ROM2@          romfetch2
S@             sfetch
END            END
```

GCD Forth Program

Listing 9.9 gcd.whp

```
\           Greatest Common Divisor

: gcd       ( x y -- gcd)
                BEGIN                       \ x y
                   OVER OVER <>             \ x y f
                WHILE                       \ x y
                   OVER OVER <              \ x y f
                   IF                       \ x y
                      OVER -                \ x y'
                   ELSE
                      TUCK - SWAP           \ x' y
                   THEN
                REPEAT                      \ x y
                DROP ;                      \ gcd

: main      ( -- )
        BEGIN
           waitB0
           S@ DUP DIG!
           waitB0
           S@ DUP DIG!
           waitB0
           gcd DIG!
        AGAIN ;
```

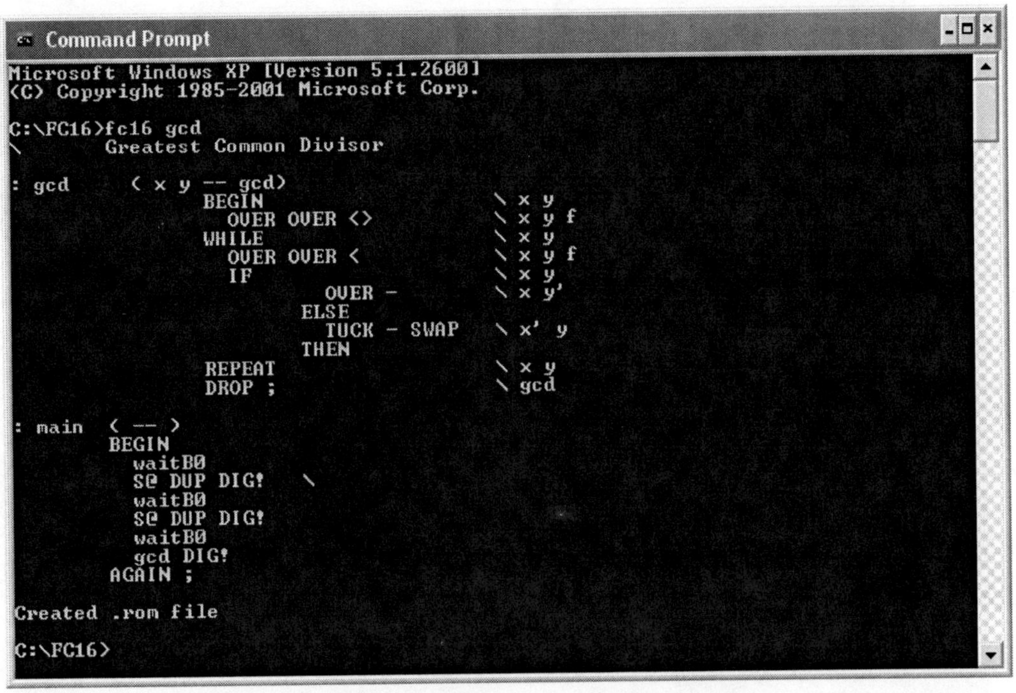

Figure 9.12 Running the FC16 compiler, *fc16.exe*

Listing 9.10 gcd.rom

```vhdl
type rom_array is array (NATURAL range <>)
                  of STD_LOGIC_VECTOR (15 downto 0);
constant rom: rom_array := (
        JMP,            --0
        X"0017",        --1
        over,           --2
        over,           --3
        neq,            --4
        JZ,             --5
        X"0015",        --6
        over,           --7
        over,           --8
        lt,             --9
        JZ,             --a
        X"0010",        --b
        over,           --c
        minus,          --d
        JMP,            --e
        X"0013",        --f
        tuck,           --10
        minus,          --11
        swap,           --12
        JMP,            --13
        X"0002",        --14
        drop,           --15
        RET,            --16
        JB0HI,          --17
        X"0017",        --18
        JB0LO,          --19
        X"0019",        --1a
        sfetch,         --1b
        dup,            --1c
        digstore,       --1d
        JB0HI,          --1e
        X"001e",        --1f
        JB0LO,          --20
        X"0020",        --21
        sfetch,         --22
        dup,            --23
        digstore,       --24
        JB0HI,          --25
        X"0025",        --26
        JB0LO,          --27
        X"0027",        --28
        CALL,           --29
        X"0002",        --2a
        digstore,       --2b
        JMP,            --2c
        X"0017",        --2d
        X"0000"         --2e
        );
```

Listing 9.11 fc16_prom_gcd.vhd

```vhdl
-- Example 53:   fc16_prom.vhd
-- gcd( x y -- gcd)
library IEEE;
use IEEE.std_logic_1164.all;
use IEEE.std_logic_unsigned.all;
use work.opcodes.all;

entity fc16_prom is
    port (
        addr: in STD_LOGIC_VECTOR (15 downto 0);
        M: out STD_LOGIC_VECTOR (15 downto 0)
    );
end fc16_prom;

architecture fc16_prom_arch of fc16_prom is
type rom_array is array (NATURAL range <>)
                  of STD_LOGIC_VECTOR (15 downto 0);
constant rom: rom_array := (
        JMP,              --0
        X"0017",          --1
        over,             --2
        over,             --3
        neq,              --4
        JZ,               --5
        X"0015",          --6
        over,             --7
        over,             --8
        lt,               --9
        JZ,               --a
        X"0010",          --b
        over,             --c
        minus,            --d
        JMP,              --e
        X"0013",          --f
        tuck,             --10
        minus,            --11
        swap,             --12
        JMP,              --13
        X"0002",          --14
        drop,             --15
        RET,              --16
        JB0HI,            --17
        X"0017",          --18
        JB0LO,            --19
        X"0019",          --1a
        sfetch,           --1b
        dup,              --1c
        digstore,         --1d
        JB0HI,            --1e
        X"001e",          --1f
        JB0LO,            --20
        X"0020",          --21
        sfetch,           --22
```

Listing 9.11 (cont.) fc16_prom_gcd.vhd

```
        dup,                    --23
        digstore,               --24
        JB0HI,                  --25
        X"0025",                --26
        JB0LO,                  --27
        X"0027",                --28
        CALL,                   --29
        X"0002",                --2a
        digstore,               --2b
        JMP,                    --2c
        X"0017",                --2d
        X"0000"                 --2e
    );

begin
  process(addr)
  variable j: integer;
  begin
      j := conv_integer(addr);
      M <= rom(j);
  end process;
end fc16_prom_arch;
```

The Aldec simulator will allow you to debug your Forth program. Fig. 9.13 shows the complete execution of the GCD calculation based on the *fc16_prom_gcd.vhd* program in Listing 9.11. The first two pressings of *btnd*(0) will push the two switch settings 0C and 08 on the data stack. The third pressing of *btnd*(0) at 710 ns will execute the word *gcd* in Listing 9.9, and this will produce the answer 0004 at 1570 ns on the signal *x*, the input to the *x7segb* component in Fig. 9.5. We used a clock frequency of 50 MHz for *clk25* in Fig. 9.13 so that the calculation of the *gcd* took

$$(1570 \text{ ns} - 710 \text{ ns})/20 \text{ ns} = 43 \text{ clock cycles}$$

You should compare this with our special purpose processor that we designed in Listing 3.1 of Example 25 in which the calculation of the GCD took only three clock cycles. (See the simulation in Fig. 3.5).

In Fig. 9.14 we have expanded the simulation to show the beginning of the GCD calculation. At 770 ns the program counter *P* is just incrementing from 0002 to 0003 and *icode* becomes the opcode for *OVER* (0004). You can compare the instructions in Listing 9.11 with the simulation in Fig. 9.14 as the program counter increases from 0002 to 000B before it jumps to 0010 because of the instruction *JZ* X"0010". You should be able to see how each instruction modifies the top two elements on the data stack, *T* and *N*, and which state the state diagram in Fig. 9.11 is in at each stage of the simulation. Note that the *exec* state is always followed by the *fetch* state and the *exec_fetch* state can persist for multiple clock cycles.

Table 9.9 compares the FPGA resources needed for the Forth core with those for the GCD special pupose processor of Example 25. The number of slices used for the *gcd* FC16 core represents 13% of the slices in the Spartan3E-500 FPGA.

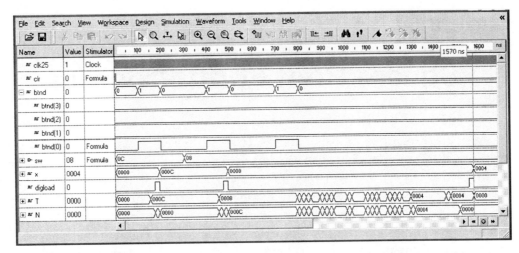

Figure 9.13 Simulation of *fc16_top.vhd* showing complete GCD calculation

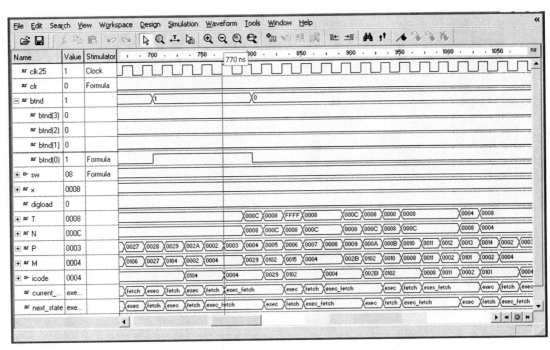

Figure 9.14 Simulation of *fc16_top.vhd* showing the beginning of the GCD calculation

Table 9.9 Comparison of FPGA resource usage

	Example 25 gcd3	Example 53 fc16_gcd_top	Example 53 fc16_prom_gcd
Total number of slices	14	651	14
Total number if 4-input LUTs	28	1,235	21
Total number of slice flip-flops	8	151	
Total number of IOB flip-flops	6	12	
Total equivalent gate count	283	17,059	189
Maximum pin delay	3.075 ns	5.593 ns	1.509 ns

Example 54

Square Root Forth Program

As a final example of using the FC16 Forth core we will compile the integer square root Forth program shown in Listing 9.12, which will produce the *sqrt.rom* file shown in Listing 9.13. We will then copy and paste this VHDL code into the same *fc16_prom* VHDL program we used in Example 53 shown in Listing 9.11. We will rename this file *fc16_prom_sqrt.vhd* but always keep the entity name as *fc16_prom*. The top-level design will be modified as shown in Listing 9.14 by adding the binary-to-BCD converter so that the results will be displayed in decimal as they were in Example 25.

Listing 9.12 sqrt.whp
```
\          Integer square root

: sqrt     ( a -- a')
                3 1 ROT              \ d s a
                BEGIN                \ d s a
                   OVER OVER <=      \ d s a f
                WHILE                \ d s a
                  -ROT OVER +        \ a d s'
                  SWAP 2 +           \ a s' d'
                  SWAP ROT           \ d' s' a
                REPEAT               \ d s a
                DROP DROP            \ d
                2/ 1- ;              \ sqrt

: main     ( -- )
      BEGIN
         waitB0
         S@ DUP DIG!
         waitB0
         sqrt DIG!
      AGAIN ;
```

The simulation shown in Fig. 9.15 shows the complete execution of the *sqrt* calculation in Listing 9.13. The first pressing of *btnd*(0) will push the switch setting 64 on the data stack and display it on the 7-segment display. The second pressing of *btnd*(0) at 410 ns will execute the word *sqrt* in Listing 9.12, and this will produce the answer 0008 at 3370 ns. We used a clock frequency of 50 MHz for *clk25* in Fig. 9.15 so that the calculation of the *sqrt* took

$$(3370 \text{ ns} - 410 \text{ ns})/20 \text{ ns} = 148 \text{ clock cycles}$$

You should compare this with our special purpose processor that we designed in Listing 3.3 of Example 26 in which the calculation of the *sqrt* took only nine clock cycles. (See the simulation in Fig. 3.8).

Listing 9.13 sqrt.rom

```vhdl
type rom_array is array (NATURAL range <>)
                  of STD_LOGIC_VECTOR (15 downto 0);
constant rom: rom_array := (
        JMP,           --0
        X"001c",       --1
        LIT,           --2
        X"0003",       --3
        LIT,           --4
        X"0001",       --5
        rot,           --6
        over,          --7
        over,          --8
        lte,           --9
        JZ,            --a
        X"0017",       --b
        mrot,          --c
        over,          --d
        plus,          --e
        swap,          --f
        LIT,           --10
        X"0002",       --11
        plus,          --12
        swap,          --13
        rot,           --14
        JMP,           --15
        X"0007",       --16
        drop,          --17
        drop,          --18
        twoslash,      --19
        minus1,        --1a
        RET,           --1b
        JB0HI,         --1c
        X"001c",       --1d
        JB0LO,         --1e
        X"001e",       --1f
        sfetch,        --20
        dup,           --21
        digstore,      --22
        JB0HI,         --23
        X"0023",       --24
        JB0LO,         --25
        X"0025",       --26
        CALL,          --27
        X"0002",       --28
        digstore,      --29
        JMP,           --2a
        X"001c",       --2b
        X"0000"        --2c
        );
```

Listing 9.14 fc16_sqrt_top.vhd

```vhdl
-- Example 54: fc16_sqrt_top
library IEEE;
use IEEE.STD_LOGIC_1164.all;
use work.fc16_components.all;

entity fc16_sqrt_top is
    port(
        mclk : in STD_LOGIC;
        btn : in STD_LOGIC_VECTOR(3 downto 0);
        sw : in STD_LOGIC_VECTOR(7 downto 0);
        ld : out STD_LOGIC_VECTOR(7 downto 0);
        a_to_g : out STD_LOGIC_VECTOR(6 downto 0);
        an : out STD_LOGIC_VECTOR(3 downto 0)
        );
end fc16_sqrt_top;

architecture fc16_sqrt_top of fc16_sqrt_top is
signal clk25, clk190, clr, digload, ldload: std_logic;
signal x,x1,S,P,M,T,N,N2,E1,E2: std_logic_vector(15 downto 0);
signal p1: std_logic_vector(9 downto 0);
signal btnd: std_logic_vector(3 downto 0);
signal b1: std_logic_vector(7 downto 0);
begin
clr <= btn(3);
S <= X"00" & sw;
x1 <= "000000" & p1;
b1 <= T(7 downto 0);
ld <= sw;
U1 : clkdiv2
    port map(mclk => mclk, clr => clr, clk25 => clk25,
         clk190 => clk190);

U2 : debounce4
    port map(cclk => clk190, clr => clr, inp => btn, outp => btnd);

U3 : fc16
    port map(P => P, S => S, M => M, E1 => E1, E2 => E2, B => btnd,
         clr => clr, clk => clk25, digload => digload,
         ldload => ldload, T => T, N => N);

U4 : fc16_prom
    port map(addr => P, M => M);

U5 : reg
    generic map(N => 16)
    port map(load => digload, d => x1, clk => clk25, clr => clr,
          q => x);
U6 : binbcd8
    port map(b => b1, p => p1);

U7 : x7segb
    port map(x => x, cclk => clk190, clr => clr, a_to_g => a_to_g,
         an => an);

end fc16_sqrt_top;
```

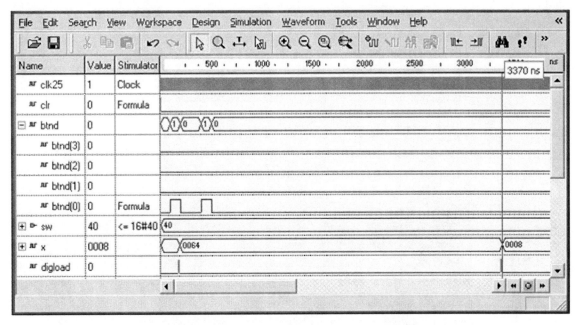

Figure 9.15 Simulation of *fc16_sqrt_top.vhd* showing complete *sqrt* calculation

In Fig. 9.16 we have expanded the simulation to show the beginning of the *sqrt* calculation. At 470 ns the program counter *P* is just incrementing from 0002 to 0003 and *icode* becomes the opcode for *LIT* (0100). Note that the value of *icode* remains 0100 for four clock cycles as 0003 and 0001 are pushed on the stack. You can compare the instructions in Listing 9.13 with the simulation in Fig. 9.16 as the program counter increases from 0002 to 0011. You should be able to see how each instruction modifies the top three elements on the data stack, *T*, *N*, and *N2*.

Table 9.10 compares the FPGA resources needed for the Forth core with those for the *sqrt* special pupose processor of Example 26 as well as the *gcd* Forth core in Example 53. The 655 slices used for the *sqrt* FC16 core represents 14% of the 4,656 slices in the Spartan3E-500 FPGA used in the Nexys-2 board. You therefore have lots of room left on the Spartan3E-500 FPGA to add more hardware, expand the FC16 Forth core, and write large Forth programs to run on the Nexys-2 board.

You could, for example, add the words "fetch" @ (*addr* -- *n*), which reads the value at address *addr* on the external RAM and puts it on the data stack, and "store" ! (*n addr* --), which stores the 16-bit value of *n* at the address *addr* on the external RAM. Because of the speed of the external RAM these instructions will take multiple 25 MHz clock cycles to execute. You would need to modify the FC16 controller by adding a clock cycle counter and then staying in the *exec* state for a specified number of clock cycles when the "fetch" or "store" opcode is executed. You could add VGA and graphics hardware from Chapters 6 and 8 and make up Forth words to plot lines and circles. This would make it easy to write interesting and sophisticated graphics programs by writing Forth program and executing them on your own personalized Forth core. Go to it!

Example 54

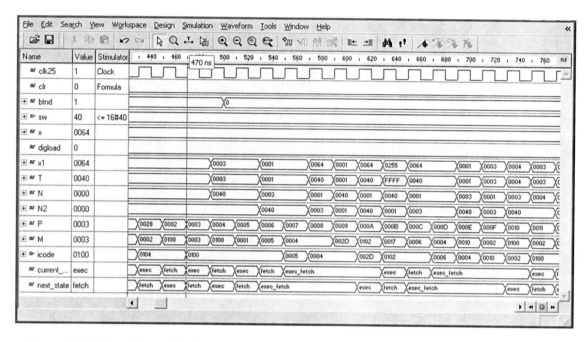

Figure 9.16 Simulation of *fc16_sqrt_top.vhd* showing the beginning of the *sqrt* calculation

Table 9.10 Comparison of FPGA resource usage

	Example 26 sqrt2	Example 54 fc16_sqrt_top	Example 53 fc16_gcd_top
Total number of slices	21	655	651
Total number if 4-input LUTs	39	1,267	1,235
Total number of slice flip-flops	14	144	151
Total number of IOB flip-flops	14	4	12
Total equivalent gate count	512	17,125	17,059
Maximum pin delay	1.941 ns	3.914 ns	5.593 ns

Appendix A

Aldec Active-HDL Tutorial – Part 1

You can download the student edition of Active-HDL from
http://www.aldec.com/education/students/
Follow the instructions and download Xilinx libraries with Active-HDL.
You can download ISE WebPACK from
http://www.xilinx.com/support/download/index.htm
Start Active-HDL by double-clicking the *avhdl.exe* icon.

Select *Create new workspace* and click *OK*.

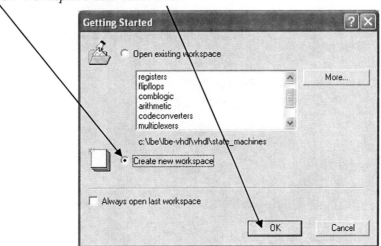

Browse to the directory where you want to store the project, type *example20* for the workspace name and click *OK*.

Select *Create an Empty Design with Design Flow* and click *Next*.

Click *Flow Settings*

Select *HDL Synthesis*

Select *Xilinx ISE/WebPack 9.1*

Press *Select*

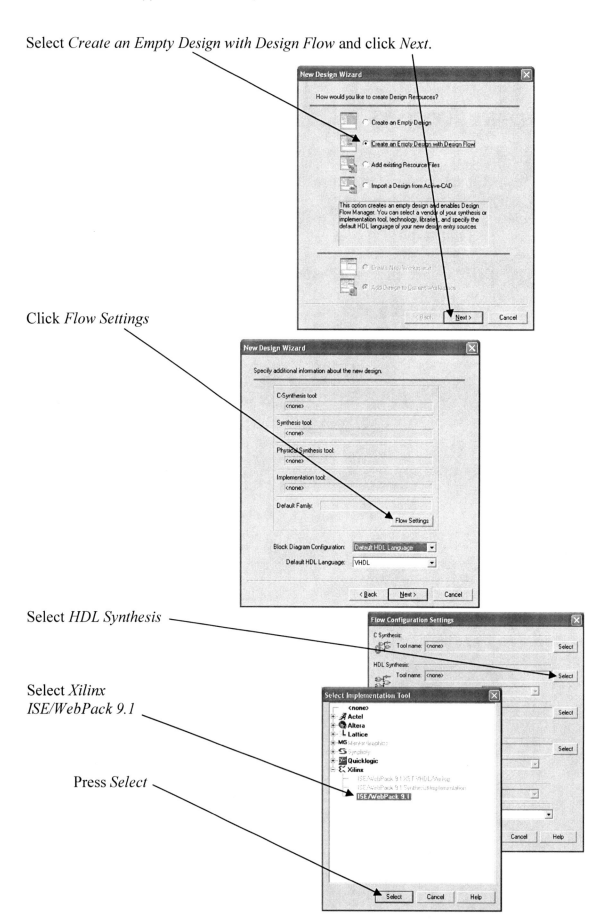

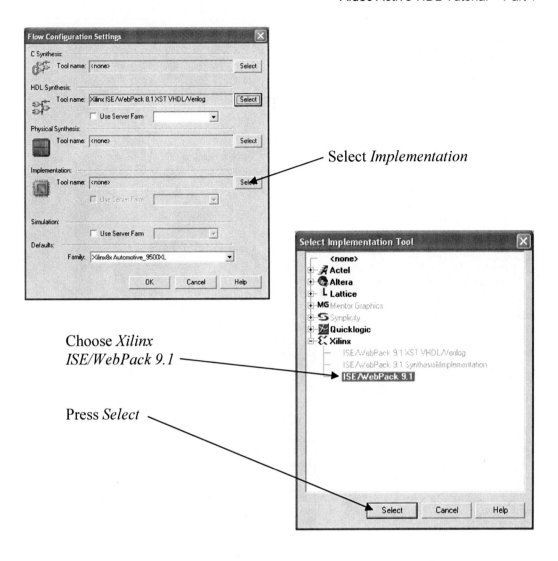

Select *Implementation*

Choose *Xilinx ISE/WebPack 9.1*

Press *Select*

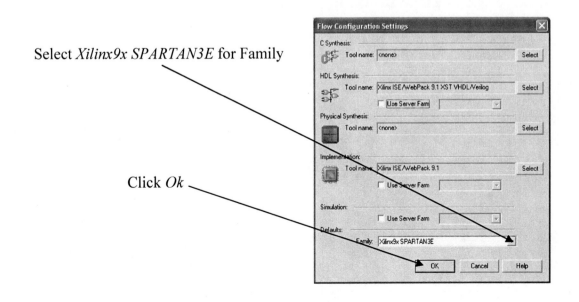

Select *Xilinx9x SPARTAN3E* for Family

Click *Ok*

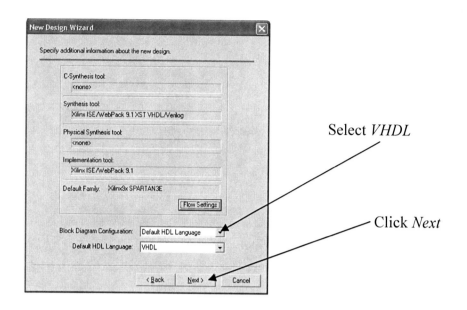

Select *VHDL*

Click *Next*

Type *doorlock* for the design name

and click *Next*.

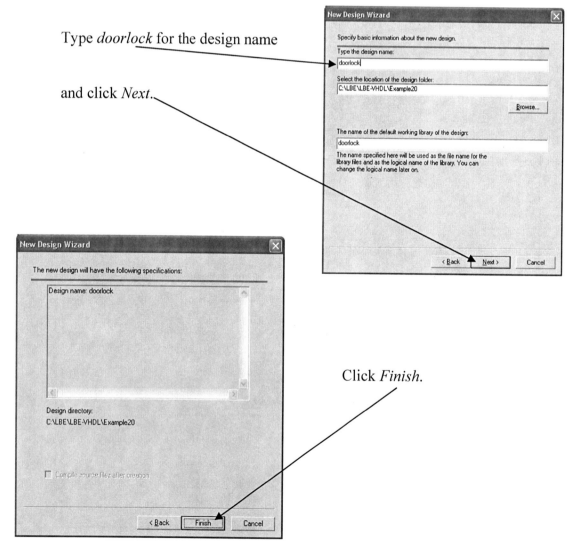

Click *Finish*.

Select *Design -> Add Files to Design...*

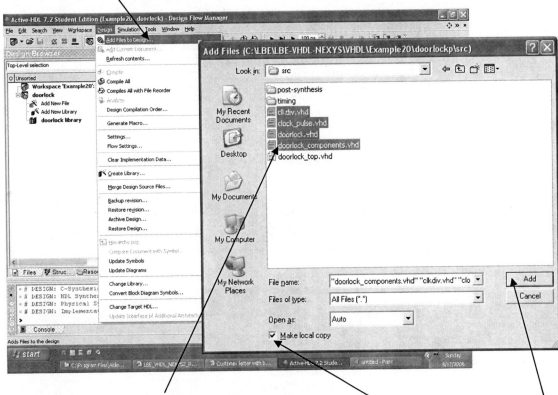

Locate the files *clkdiv.vhd*, *clock_pulse.vhd*, *doorlock.vhd*, and *doorlock_components.vhd*

Check *Make local copy*, then click *Add*.

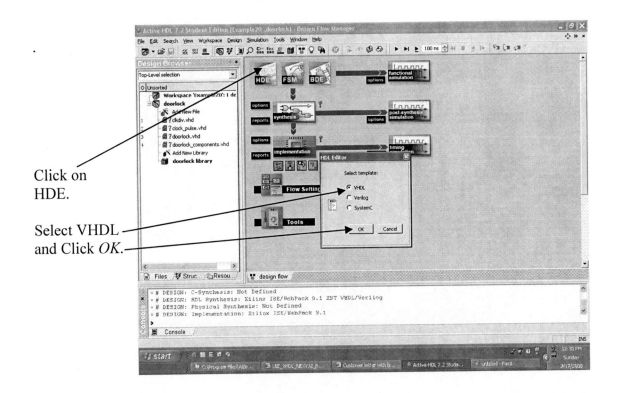

Click on HDE.

Select VHDL and Click *OK*.

Click Next.

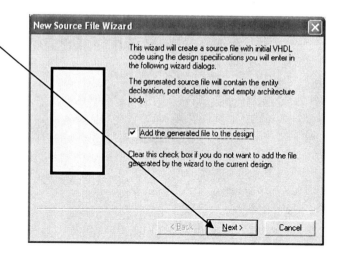

Type *doorlock2_top* and click *Next*.

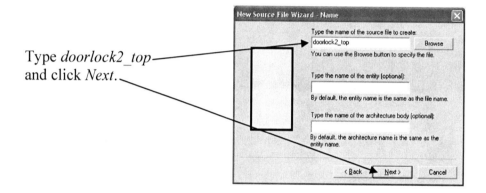

Click *New*.

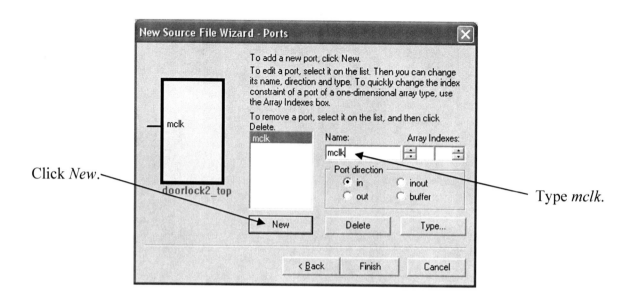

Type *mclk*.

Aldec Active-HDL Tutorial – Part 1

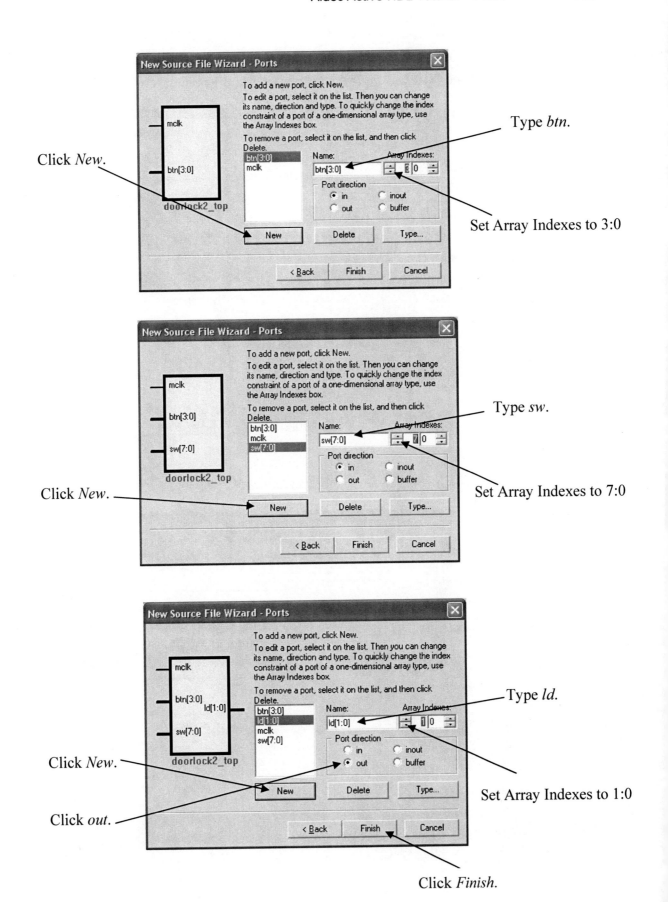

This will generate a VHDL template with the input and output signals filled in. Add your name as author and type *Doorlock code from switch settings* as the description. Delete the other comments.

Note that the **entity** has been completed for you.

1 Click *Save*

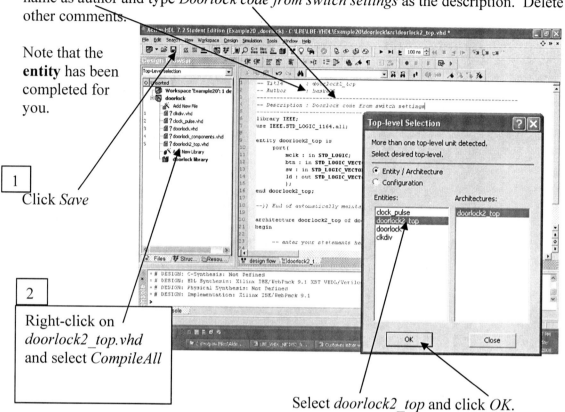

2 Right-click on *doorlock2_top.vhd* and select *CompileAll*

Select *doorlock2_top* and click *OK*.

Double-click on *doorlock_components.vhd*

Click + on *doorlock* library

Right Click on *clock_pulse* and select *Copy Declaration*

These *component declarations* can be pasted into the *doorlock_components* package as we have done here. You should delete the last *for all:* statement

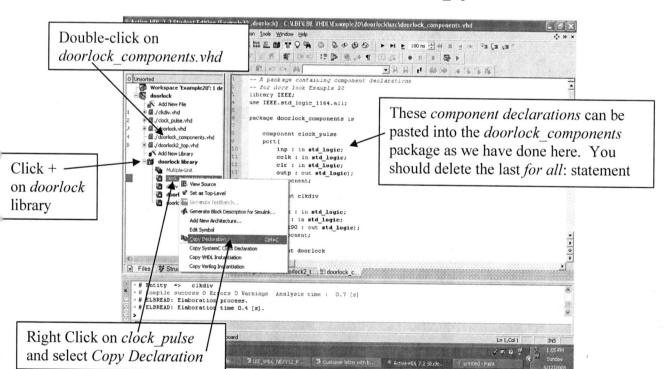

Aldec Active-HDL Tutorial – Part 1

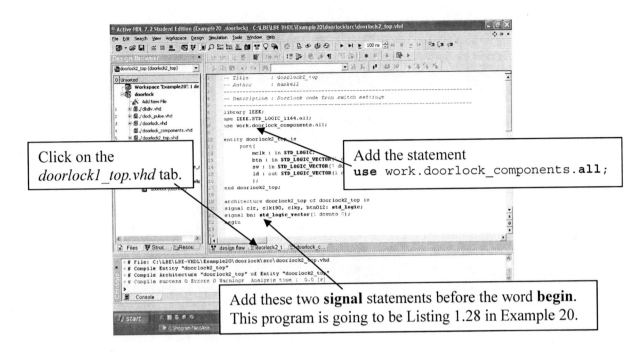

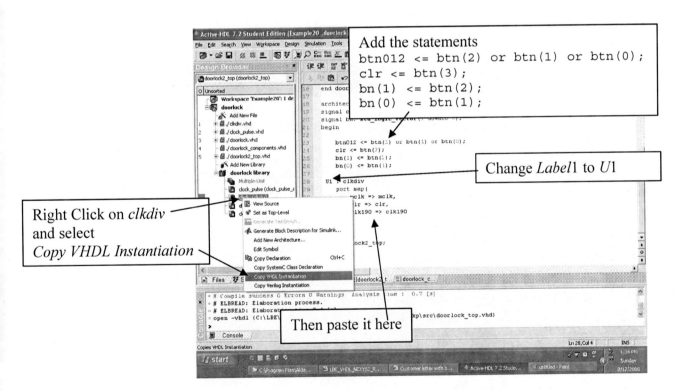

Repeat this procedure to copy the VHDL instantiations for the two components *clock_pulse*, and *doorlock*.

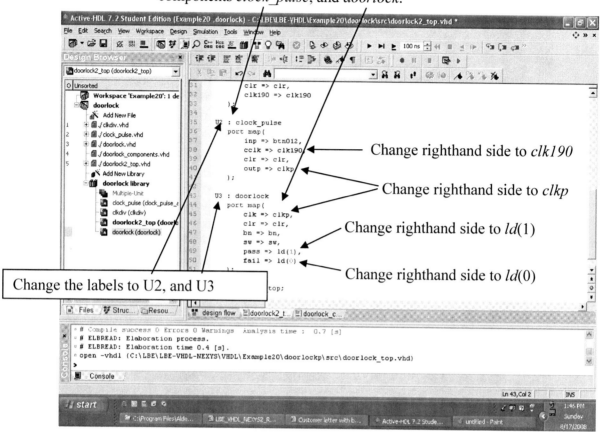

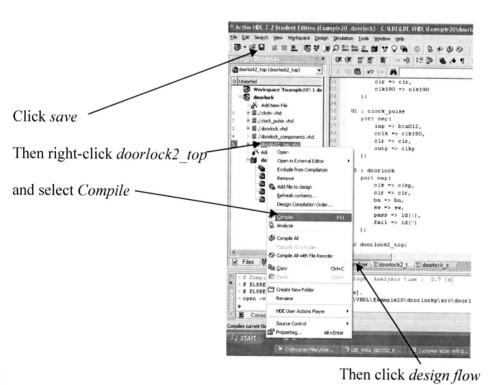

Aldec Active-HDL Tutorial – Part 1

Click *options*

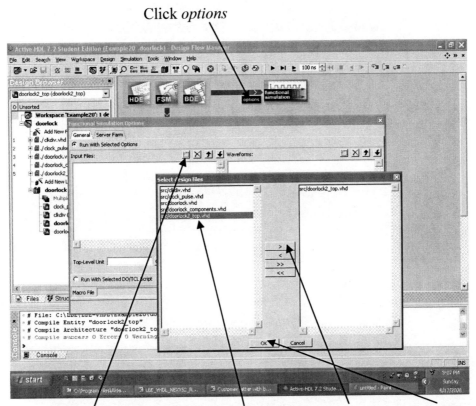

Click *here* and select *doorlock2_top.vhd*. Click >. Then click *OK*.

Click *Choose*.

Select *doorlock2_top* and click *Add*.

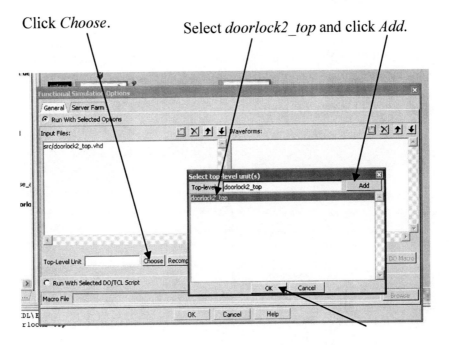

Click *OK*.

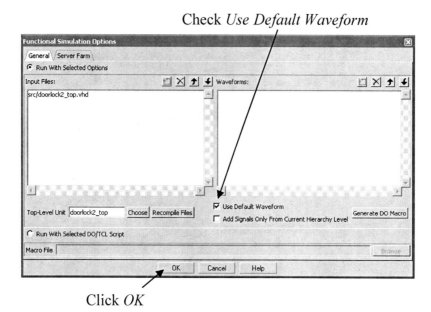

Check *Use Default Waveform*

Click *OK*

Click *functional simulation*

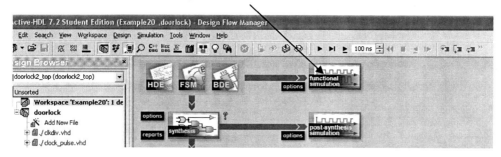

The waveform window will automatically come up with the simulation already initialized. Make sure order is *clkp*, *clr*, *sw*. *btn*, *bn*, *btn012*, *ld* (grab and drag if necessary). Delete the other signals.
Right-click on *clkp* and select *Stimulators*.

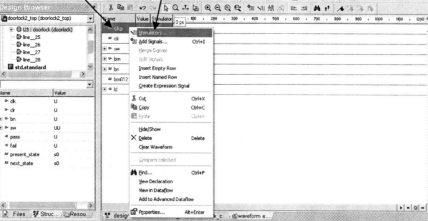

Select *Clock* and set Frequency to 50 MHz

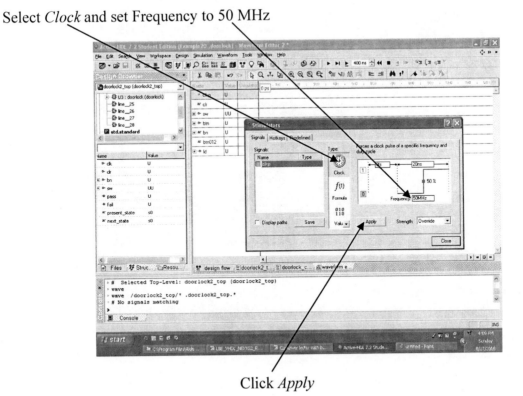

Click *Apply*

Click on *clr*, select *Formula* and set to 1 0ns, 0 10ns

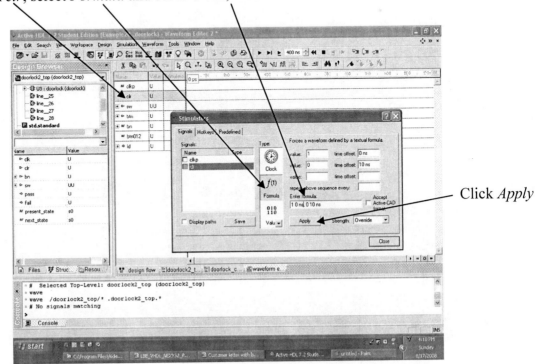

Click *Apply*

Click on *sw*, select *Value* and set to 16#18.

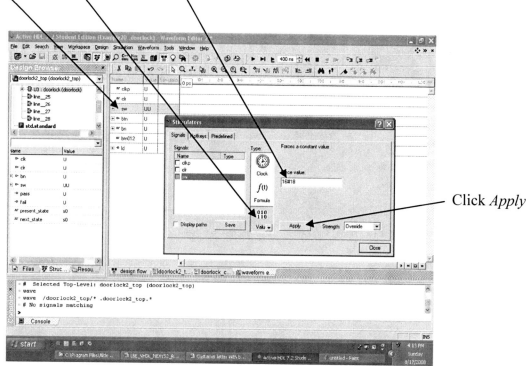

Click *Apply*

Click on *btn*, select *Counter* of type *Circular One*. Count every 20 ns.

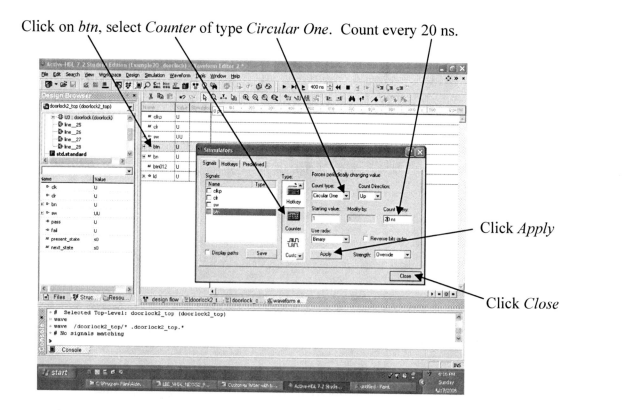

Click *Apply*

Click *Close*

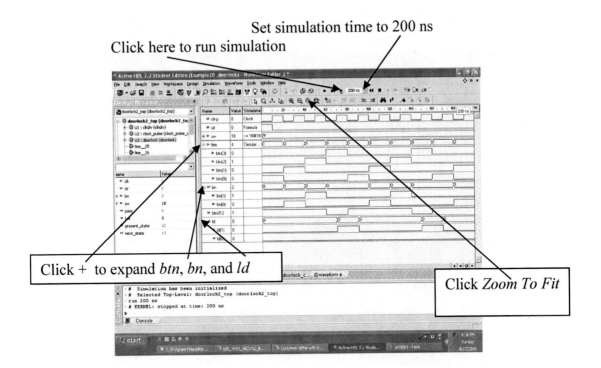

Note that pressing the button sequence 0-1-2-0 will light the pass LED *ld*(1). See Examples 19 and 20 for a discussion of how the doorlock code works.

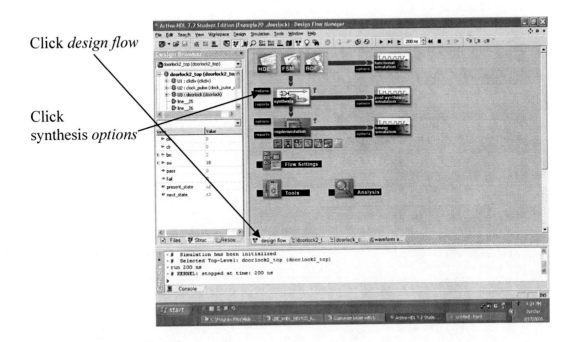

Pull down menu and select *doorlock2_top* for Top-level Unit.

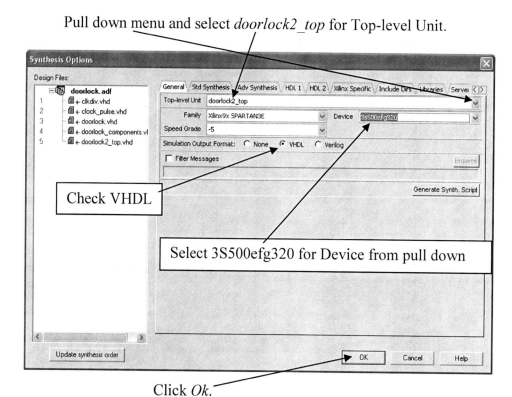

Check VHDL

Select 3S500efg320 for Device from pull down

Click *Ok*.

Click *synthesis*

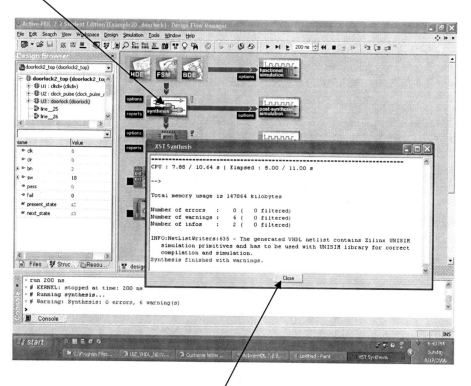

After synthesis is complete, click *Close*.

Aldec Active-HDL Tutorial – Part 1

Click implementation *options*

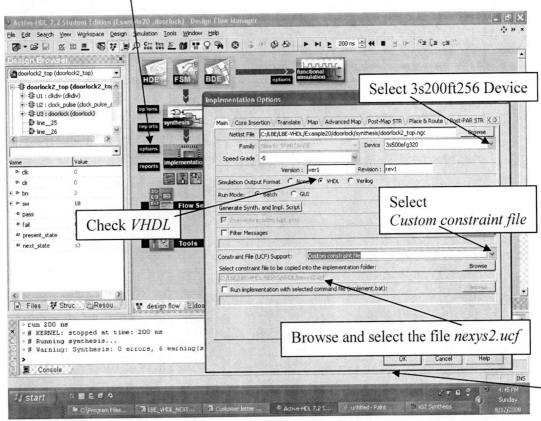

Select 3s200ft256 Device

Select *Custom constraint file*

Check *VHDL*

Browse and select the file *nexys2.ucf*

Select *Translate* and check *Allow Unmatched LOC Constraints.*

Select *BitStream* and uncheck *Do Not Run Bitgen.*

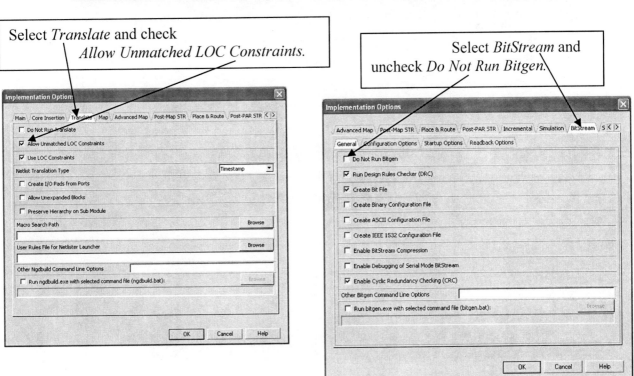

Click *Startup Options* within the *BitStream* tab and select *JTAG Clock*

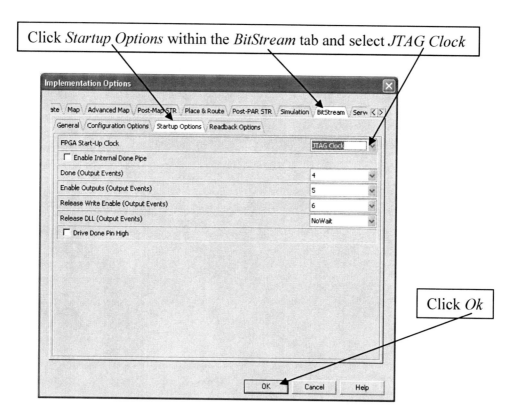

Click *Ok*

Click *implementation*

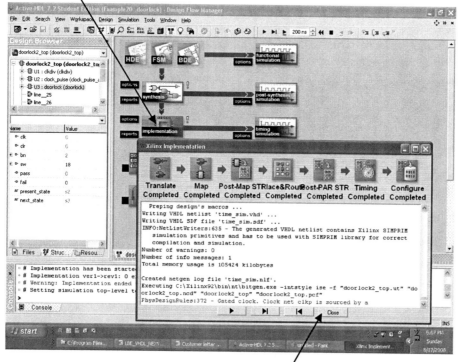

When implementation is complete click *Close*.

Program Nexys-2 Board

To program the Spartan3E on the Nexys-2 board we will use the **ExPort** tool that is part of the the **Adept Suite** available free from Digilent at http://www.digilentinc.com/Software/Adept.cfm?Nav1=Software&Nav2=Adept Connect the Nexys-2 board to your computer using the USB cable. Make sure the power select jumper JP7 is set to USB. Double-click the **ExPort** icon on the desktop.

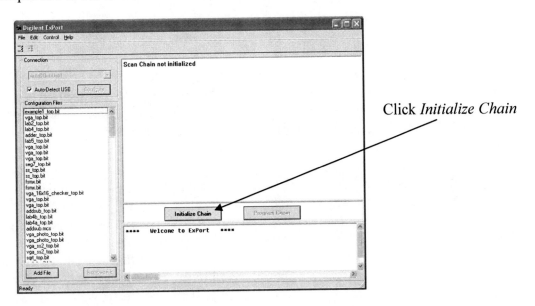

Click *Initialize Chain*

Click *Browse* and go to *Example20* → *doorlock* → *implement* → *ver1* → *rev1*
Select *doorlock2_top.bit*

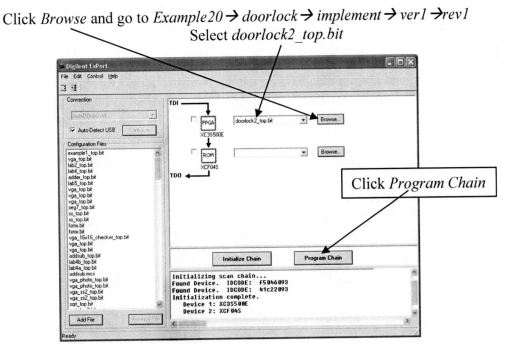

Click *Program Chain*

Your program is now running on the board. Set the switches to the doorlock code you want (see Example 19 in text) and verify its operation by pressing the proper buttons.

Appendix B

Test Bench and XPower Tutorial

Performing a Timing Simulation

Introduction

In Example 24, we created a special-purpose processor to compute the integer square root of an input. Figure 1.12 shows the results from performing a functional simulation to obtain the integer square root of 64. Until now we have performed functional simulations to verify that our design produces the results expected. Functional simulations are completely independent of the target FPGA (or other chip). After synthesizing and implementing the design to a particular target FPGA, it is important to verify that the design will still work correctly on the target. This simulation, called a *timing simulation*, takes into account parameters specific to the implemented design including timing given the delays caused by the way the particular design was placed & routed on the FPGA. To thoroughly test our square root component, we should create a test bench with many different inputs and outputs to ensure that the circuit produces the expected results in as many cases as possible. Initially, we would like to see if the timing simulation for at least one or a few test cases will run properly. For this, we can use ActiveHDL to automatically generate a basic test bench from the stimulators that we set for *clk, clr, strt,* and *sw* for the functional simulation shown in Fig. 2.12.[5]

Preparing the Functional Simulation

Repeat the functional simulation from Example 24 with the following stimulators: *clr* =<*formula*> 1 0 ns, 0 5 ns, *clk* =<*clock*> 50 MHz, *strt* =<*formula*> 0 0 ns, 1 40 ns, 0 120 ns; *sw* =<*value*> 16#40. Be sure to run the simulation for 500 ns. Even though the answer appears at approximately 410 ns, we will notice later that once timing delays from the FPGA are added, it will take longer to compute the answer. Save this functional simulation as *sqrt.awf* using the File → Save selection from the menu bar.

---

[5] This tutorial requires the full version of Active-HDL and can not be completed with the Student version.

Creating a Basic Test Bench from the Waveform for the Functional Simulation

After synthesizing and implementing the top level, *sqrt*, additional libraries will appear in the ActiveHDL workspace including additional folders containing a post-synthesis top level and some time simulation files. Expand the *sqrt library* by clicking on the '+' sign next to the library.

Right-click on the top-level, *sqrt*, component within that library for more options.

Finally, select *Generate TestBench* to launch a wizard that will create a testbench using the waveform from the functional simulation.

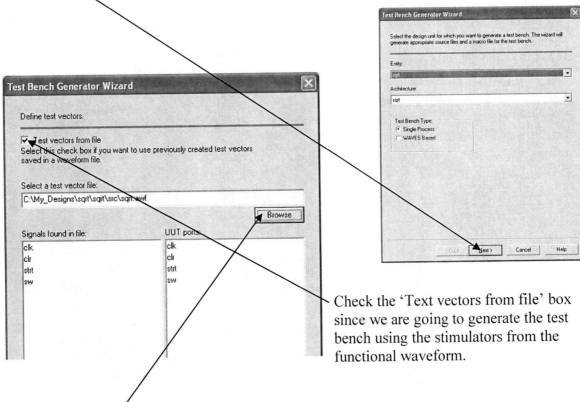

The first step in the wizard contains the entity and architecture for the design's top level. Click *Next*.

Check the 'Text vectors from file' box since we are going to generate the test bench using the stimulators from the functional waveform.

Next, use the *Browse* button to search for and select the waveform that was saved from the functional simulation. Click *Next*.

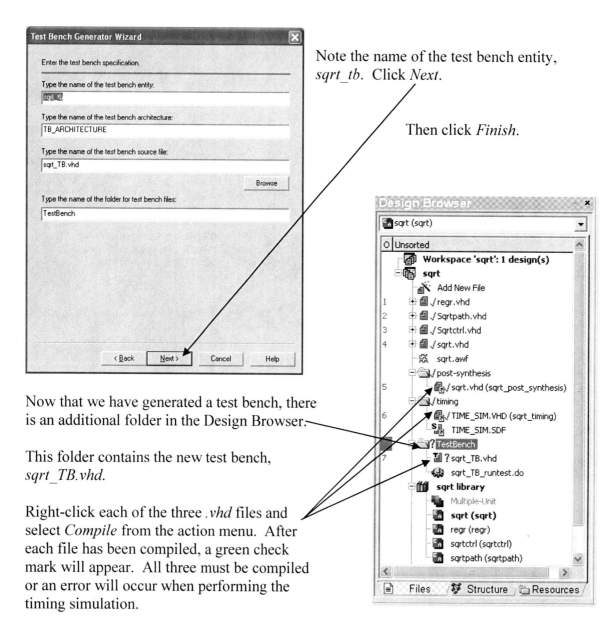

Note the name of the test bench entity, *sqrt_tb*. Click *Next*.

Then click *Finish*.

Now that we have generated a test bench, there is an additional folder in the Design Browser.

This folder contains the new test bench, *sqrt_TB.vhd*.

Right-click each of the three *.vhd* files and select *Compile* from the action menu. After each file has been compiled, a green check mark will appear. All three must be compiled or an error will occur when performing the timing simulation.

Double-click the *sqrt_tb.vhd* file to see the test bench that was generated.

```
entity sqrt_tb is
end sqrt_tb;

architecture TB_ARCHITECTURE of sqrt_tb is
    -- Component declaration of the tested unit
    component sqrt
    port(
        clk : in std_logic;
        clr : in std_logic;
        strt : in std_logic;
        sw : in std_logic_vector(7 downto 0);
        root : out std_logic_vector(3 downto 0) );
    end component;
```

Basically, the test bench is a wrapper that wraps around our original top-level component and controls the input signals according to the stimulus. Therefore, the actual top-level entity does not have any inputs or outputs because the inputs are set by internal signals created by the stimulators from the functional waveform. The output will be viewed on the timing simulation waveform.

Stimulator signals have been automatically created.

The *sqrt* component is port mapped to these signals.

```
signal clk : std_logic;
signal clr : std_logic;
signal strt : std_logic;
signal sw : std_logic_vector(7 downto 0);
-- Observed signals - signals mapped to the output ports
signal root : std_logic_vector(3 downto 0);

--Signal is used to stop clock signal generators
signal END_SIM: BOOLEAN:=FALSE;

-- Add your code here ...

begin

    -- Unit Under Test port map
    UUT : sqrt
        port map (
            clk => clk,
            clr => clr,
            strt => strt,
            sw => sw,
            root => root
        );
```

A *STIMULUS* process contains statements that set these stimulus signals according to the functional waveform. Compare the statements/timing in the VHDL code with the waveform.

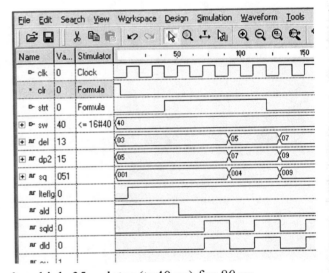

```
STIMULUS: process
begin  -- of stimulus process
--wait for <time to next event>; -- <current time>

    clr <= '1';
    strt <= '0';
    sw <= "01000000";
    wait for 5 ns;  --0 fs
    clr <= '0';
    wait for 35 ns;  --5 ns
    strt <= '1';
    wait for 80 ns;  --40 ns
    strt <= '0';
    wait for 380 ns;  --120 ns
    END_SIM <= TRUE;
    -- end of stimulus events
    wait;
end process; -- end of stimulus process
```

After *5 ns clr* changes from high to low. *Strt* pulses high *35 ns* later (t=40 ns) for *80 ns* (until t=120 ns). Finally, the simulation ends *380 ns* after that (at t=500 ns).

Finally, the *CLOCK_clk* process generates the clock signal from the clock stimulator in the functional waveform. If the *END_SIM* signal is false, the clock signal oscillates between 0 and 1 every *10 ns*. If the *END_SIM* signal is high, the clock stops and the simulation is halted. The simulation will last the same length of time as the simulation performed in the saved functional waveform.

```
CLOCK_clk : process
begin
    --this process was generated based on .
    --wait for <time to next event>; -- <c
    if END_SIM = FALSE then
        clk <= '0';
        wait for 10 ns;  --0 fs
    else
        wait;
    end if;
    if END_SIM = FALSE then
        clk <= '1';
        wait for 10 ns;  --10 ns
    else
        wait;
    end if;
end process;
```

Next, select the *Options* button for the Timing Simulation from the Flow Menu.

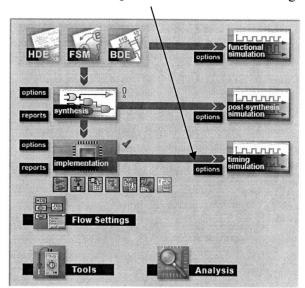

Remove the file listed in the *Input Files* box by selecting the file and clicking the 'X' icon.

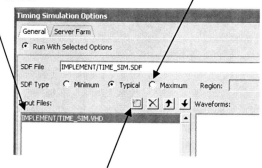

Click the icon for adding files to add the correct file for the new test bench.

Select the *sqrt_TB.vhd* file and click *OK*.

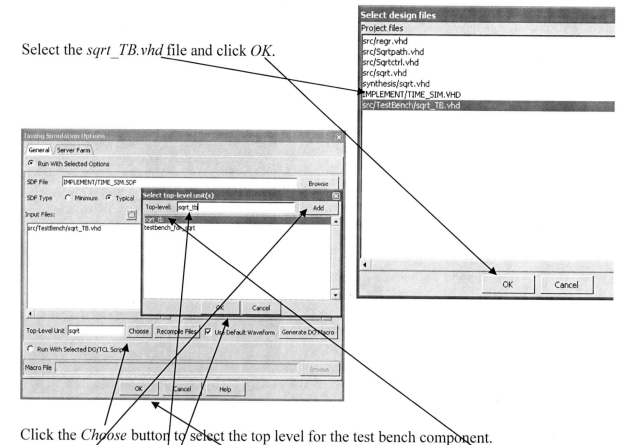

Click the *Choose* button to select the top level for the test bench component.

Be sure to delete the *sqrt* top level out of the top-level text box then click *sqrt_tb* and click *Add*.

Finally, click *OK*.

Then, Click *OK* on the timing simulation options dialog box.

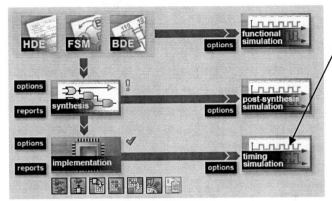

Now we are ready to run the timing simulation. Click the *Timing Simulation* button from the flow menu.

Click the run icon to run the simulation for 500 ns.

Note that although the result appeared at approximately 410 ns on the waveform from the functional simulation, it did not appear until approximately 475 ns on this timing simulation waveform. Once timing information is included, execution requires slightly more time than estimated by the target-independent functional simulation.

Save this timing simulation waveform using File → Save as *sqrt_time.awf*.

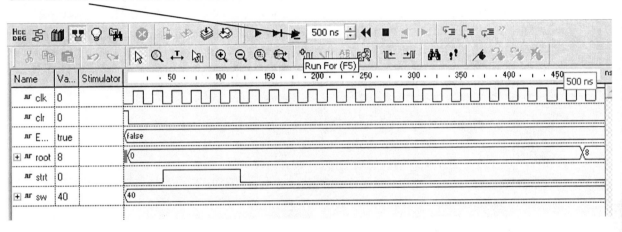

Timing Simulation

Functional Simulation – No Timing

These figures show zooming in to when the result appears on both the timing and functional waveforms. On the functional simulation the result appears exactly in-sync with the clock pulse. However, the timing waveform shows the realistic delay from when the clock goes high on that final cycle and when the output is actually valid!

Calculating the Power Required

Power involves several different factors. First, there is a base power for the FPGA, even if it is inactive. In addition to the base power, the clock requires power and the more frequently signals internal to the FPGA switch, the more power it requires. The power required by the clock is directly related to its frequency; as the clock frequency increases, the power required increases.

ActiveHDL links to a tool, XPower, created by Xilinx to calculate and measure the power required for a design. In this tutorial, we will use XPower to explore the power required to calculate the integer square root of 64 using our Integer Square Root component. Open the ActiveHDL project that contains the workspace for the Integer Square Root component. Also, in order to complete this tutorial, the timing simulation waveform, *sqrt_time.awf*, from the previous section is required. To perform a timing simulation, refer to the previous section. Synthesize and Implement the *sqrt* top level before beginning.

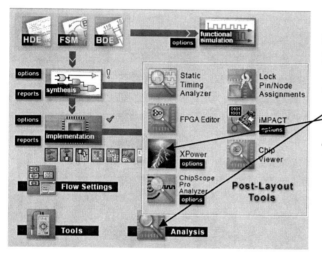

Click on the *Analysis* option on the flow menu.

Then, click the *XPower* button.

This will launch the XPower tool.

The summary tab shows the total power required by the idle FPGA, 37.00 mW.

In order to see any difference, we must first set a clock frequency.

Select 'Tools' → 'Estimate Activity Rates' from the menu bar.

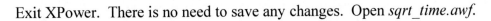

After selecting to Estimate Activity Rates, we are presented with all of the input signals for our design. Enter *50.00* in the frequency for the *clk* signal and press Enter. Note that the power is updated to 38.11 mW. This means that with an input clock of 50 MHz and no other signals changing, the design requires 1.11 mW more than the idly 37.00 mW with no activity. At this point, these are only estimates. Since the exact power depends on how *every* signal in the FPGA changes, it would be necessary to provide XPower with a simulation that contains every signal change over time for a particular execution. Recall that the timing simulation contains these details; it is specific to the target FPGA. In this tutorial we wish to determine the exact power required to compute the integer square root of 64, so we will use the timing simulation waveform, *sqrt_time.awf* to create a *variable change dump (.VCD)* file. A variable change dump file is used as input to XPower. This file is written in a special scripting language that identifies when every signal in an FPGA changes. ActiveHDL can export a timing simulation waveform to a .VCD file.

Exit XPower. There is no need to save any changes. Open *sqrt_time.awf*.

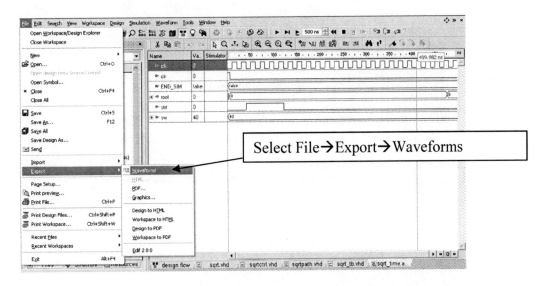

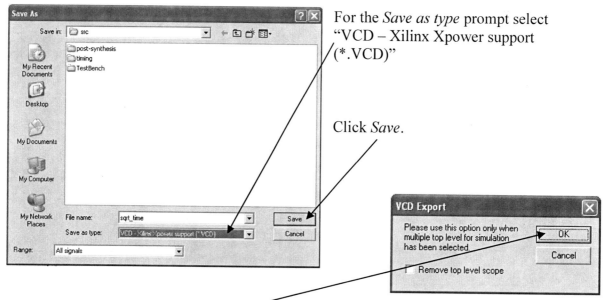

For the *Save as type* prompt select "VCD – Xilinx Xpower support (*.VCD)"

Click *Save*.

A warning will appear. Click *OK*.

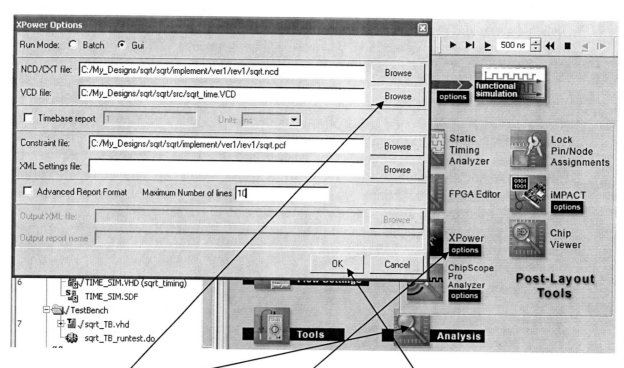

From the flow menu, click *Analysis*, then *Options* next to XPower.

Use the 'Browse' button on the XPower Options dialog box to search for the *sqrt_time.VCD* file that we exported. Usually, this is in the *src* directory. Finally, click *OK*.

Now that we have set the .VCD file, click the XPower icon from the Analysis submenu on the flow menu. This will launch XPower.

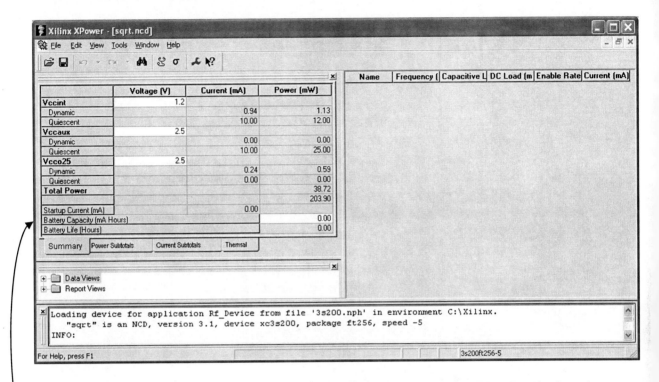

Notice that the total power is 38.72 mW. Recall that the design with a 50 MHz clock and nothing else calculated to 38.11 mW of power. Once we added all of the additional information, an actual power calculation was made and showed that it requires 38.72 mW to compute the integer square root of 64. This can also calculate battery life given the battery's capacity.

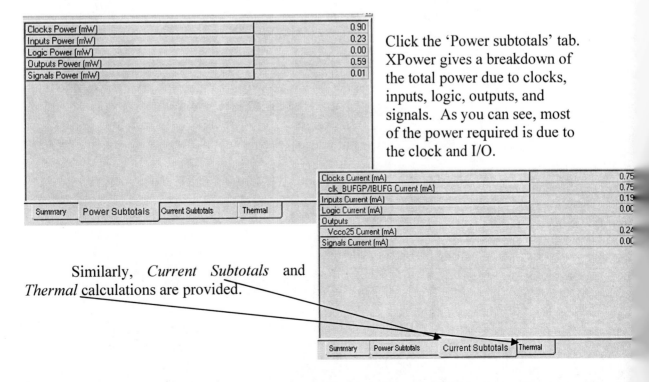

Click the 'Power subtotals' tab. XPower gives a breakdown of the total power due to clocks, inputs, logic, outputs, and signals. As you can see, most of the power required is due to the clock and I/O.

Similarly, *Current Subtotals* and *Thermal* calculations are provided.

Appendix C

Making a Turnkey System

To program the FPGA you have been downloading a *.bit* file using the ExPort program from Digilent. This configures the RAM bits in the FPGA to implement your particular design. When you turn the power off, your design goes away.

To make a turnkey system that will activate your hardware every time power is applied to the board, you must store the bit file in an onboard program PROM. When you set the MODE jumper JP9 on the board to the RUN mode then the bit file stored in the PROM will automatically by loaded into the FPGA when power is applied to the board.

To program the PROM you must create a *.mcs* file from you *.bit* file. You can do this by running iMPACT, which comes with the Xilinx ISE WebPACT.

Run iMPACT and check *create a new project*.

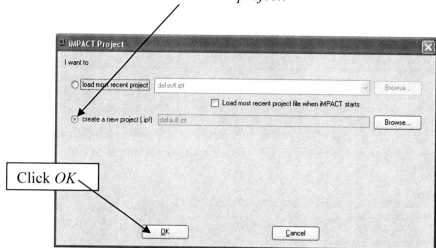

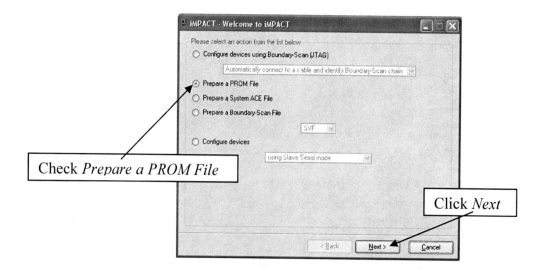

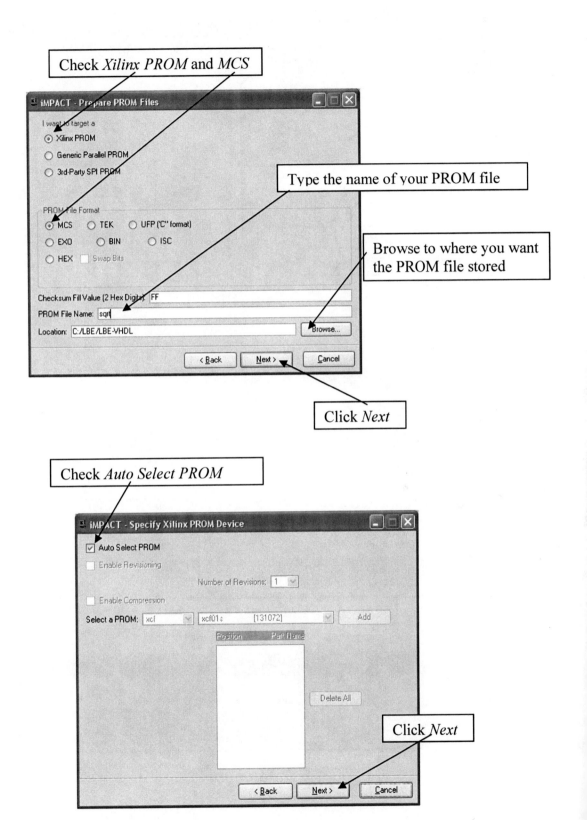

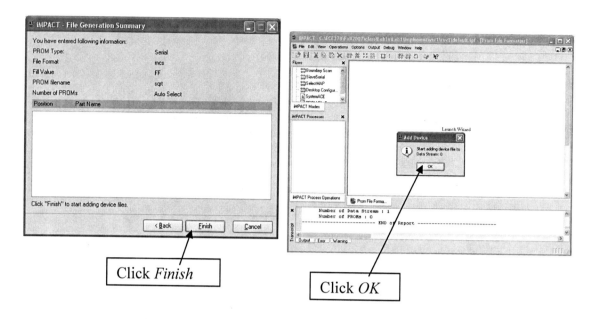

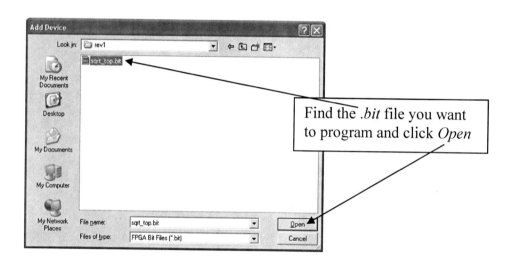

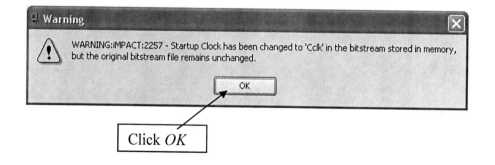

Making a Turnkey System

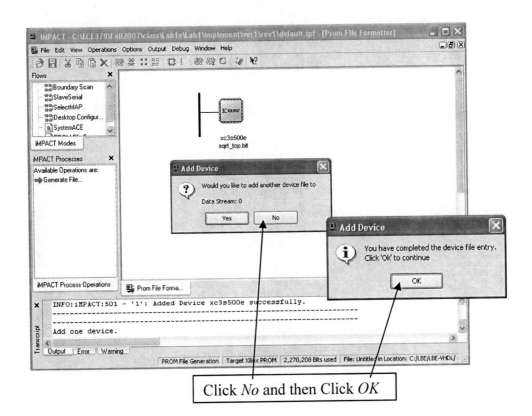

Click *No* and then Click *OK*

Double-click *Generate File*

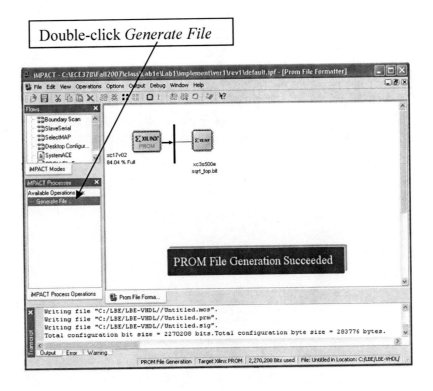

Your *.mcs* PROM file has now been generated and stored in the location you specifed. Run ExPort, *Initialize Chain*, and *check* to bypass the FPGA.

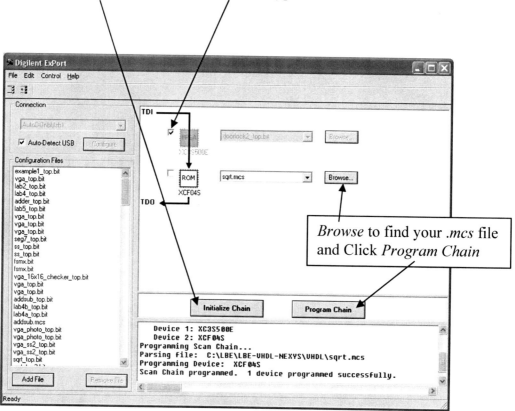

The program PROM on the Nexys-2 board is now programmed with your design. Remove power from the board. Change the MODE jumper JP9 on the board to the RUN mode. Apply power to the board and your program will be running.

Appendix D

VHDL Quick Reference Guide

Category	Definition	Example

Identifer Names	Can contain any letter, digit, or underscore _
	Must start with alphabetic letter
	Can not end with underscore or be a keyword
	Case insensitive
		`q0`
		`Prime_number`
		`lteflg`

Signal Values	'0' = logic value 0 '1' = logic value 1 'Z' = high impedance 'X' = unknown value	

Numbers and Bit Strings	\<base\>#xxx#
	B = binary
	X = hexadecimal
	O = octal
		`35 (default decimal)`
		`16#C# = "1100"`
		`X"3C" = B"00111100"`
		`O"234" = B"010011100"`

Generic statement	Associates an identifer name with a value that can be overridden with the **generic map** statement	`generic ( N:integer := 8);`
generic map	Assigns a value to a generic parameter	`generic map (N => 16)`

Signals and Variables Types	**signal** (used to connect one logic element to another)
	variable (variables assigned values in process)
	integer (useful for loop control variables)
		`signal d : std_logic_vector(0 to 3);`
		`signal led: std_logic;`
		`variable q: std_logic_vector(7 downto 0);`
		`variable k: integer;`

Program structure	`library IEEE;`
	`use IEEE.STD_LOGIC_1164.all;`
	
	`entity <identifier> is`
	`    port(`
	`        <port interface list );`
	`end <identifier>;`
	
	`architecture <identifier> of`
	`              <entity_name> is`
	`begin`
	`  process(clk, clr)`
	`  begin`
	`     {{concurrent_statement}}`
	`  end<identifier>;`
		`library IEEE;`
		`use IEEE.STD_LOGIC_1164.all;`
		
		`entity Dff is`
		`   port(`
		`       clk : in STD_LOGIC;`
		`       clr : in STD_LOGIC;`
		`       D : in STD_LOGIC;`
		`       q : out STD_LOGIC );`
		`end Dff;`
		
		`architecture Dff of Dff is`
		`begin`
		`  process(clk, clr)`
		`  begin`
		`     if(clr = '1') then`
		`        q <= '0';`
		`     elsif(rising_edge(clk))then`
		`        q <= D;`
		`     end if;`
		`  end process;`
		`end Dff;`

Logic operators	not
	and
	or
	nand
	nor
	xor
	xnor
		`z <= not y;`
		`c <= a and b;`
		`z <= x or y;`
		`w <= u nand v;`
		`r <= s nor t;`
		`z <= x xor y;`
		`d <= a xnor b;`

VHDL Quick Reference Guide (cont.)

Arithmetic operators	+ (addition) - (subtraction) * (multiplication) / (division) (not synthesizable) rem (remainder)	`count <= count + 1;` `q <= q - 1;`
Relational operators	=, /=, >, <, >=, <=	`if a <= b then` `if clr = '1' then`
Shift operators	shl (arg,count) shr (arg,count)	`c = shl(a,3);` `c = shr(a,4);`

process	[<id>] **process**(<sensitivity list>)
	{{process declaration}}
	begin
	  {{sequential statement}}
	end process [<id>]
		`process(a)`
		  `variable j: integer;`
		`begin`
		  `j := conv_integer(a);`
		  `for i in 0 to 7 loop`
		    `if(i = j) then`
		      `y(i) <= '1';`
		    `else`
		      `y(i) <= '0';`
		    `end if;`
		  `end loop;`
		`end process;`

if statement	**if**(expression1) **then**
	  {{statement;}}
	{{**elsif** (expression2) **then**
	  {{statement;}} }}
	[[**else**
	  {{statement;}}]]
	end if;
		`if(clr = '1') then`
		  `q <= '0';`
		`elsif(clk'event and clk = '1') then`
		  `q <= D;`
		`end if;`

case statement	**case** expression **is**
	((**when** choices => {sequential
	                  statement;}}))
	{{ ... }}
	when others => {sequential
	                 statement;}}
	end case;
		`case s is`
		  `when "00"  => z <= c(0);`
		  `when "01"  => z <= c(1);`
		  `when "10"  => z <= c(2);`
		  `when "11"  => z <= c(3);`
		  `when others => z <= c(0);`
		`end case;`

for loop	**for** identifier **in** range **loop**
	  {{sequential statement}}
	end loop;
		`zv := x(1);`
		`for i in 2 to 4 loop`
		  `zv := zv and x(i);`
		`end loop;`
		`z <= zv;`

Assignment operator	:= (variable) <= (signal)	`z := z + x(i);` `count <= count + 1;`

Port map	instance_name component_name **port map**
	(port_association_list);
		`M1 : mux21a port map(`
		  `a => c(0), b => c(1),`
		  `s => s(0), y => v);`

Index

Active-HDL, (*see* Aldec)
Aldec:
    Active-HDL, 3
    Active-HDL tutorial, 313-41
Algorithms:
    circle plotting (*see* graphics)
    GCD, 66-67, 91-94, 301-7
    line plotting (*see* Bresenham's
        line algorithm)
    square root, 77-88, 95-98, 308-12
ALU, 22-25
Arithmetic Logic Unit (*see* ALU)
ASCII codes, 131

Binary-to-BCD conversion, 27-30
Bresenham's line algorithm, 228-32

Circle, (*see* graphics)
Clock divider, 14-16
CLB, 1-2
Clock pulse, 11-12
Comparator, 16-17
    4-bit using relational operators,
        16-17
Configurable logic block, (*see* CLB)
Control unit, 64-74, 81-83, 89
Conversions:
    binary-to-BCD, 27-30
    binary-to-Gray code, 30-32
    Gray code to binary, 30, 33
Core generator, 104-8
Counters, 13-14
    N-bit, 13-14

D/A converter, 146-47
Datapath, 64-74, 78, 89
Debounce, 10-11
Decoder, 40-41
Digilent, 3
Discriminating function, 252-53
Divider, 38-39

Door lock code, 48-56

FC16 (*see* Forth)
    compiler, 301
Field programmable gate array, (*see*
    FPGA)
Finite state machine, (*see*
    state machine)
Flash memory, (*see* memory)
Forth, 264-312
    arithmetic and logical words,
        266-67
    branching and looping, 267, 292
    conditional words, 267
    core, 272-79
    data stack, 277, 280-82
    data stack operations, 281
    engines, 271-72
    FC16 controller, 291-300
    functional unit, 283-88
    opcodes, 274-75
    programming language, 265-67
    return stack, 289-90
    stack manipulation words, 266
    words, 266-68, 274-75, 281, 283,
        290, 292
    writing programs, 269-71
FPGA, 1-2
    Spartan-3E, 1-2

Gates:
    4-input, 25-27
GCD (*see* algorithms)
    Forth program, 301-7
Graphics, 215-63
    clearing the screen, 216-21
    plotting a circle, 250-63
    plotting a dot, 222-27
    plotting a line, 228-41
    plotting a star, 242-49
Gray code, 30
    conversions, 30-33

Keyboard, 191-94
    scan codes, 192

Line plotting (*see* graphics)
Lookup table (*see* LUT)
LUT, 1

Memory, 99-129
    flash, 126-29
    RAM (*see* RAM)
    ROM (*see* ROM)
    stack (*see* stack)
    video flash, 182-87
Moore, Chuck, 265
Mouse, 195-214
Multiplexer, 5-6, 18-19
    2-to-1, 5-6
    4-to-1, 18-19
    generic, 5
Multiplier, 34-37
MUX, (*see* multiplexer)

Nexys-2 board, 1, 3

Power, 338-41
PS/2 port, 188-214

RAM:
    block, 116-18
    distributed, 104-8
    external, 119-25
    external video, 175-81, 216
Register, 8-9
    N-bit, 8-9
ROM, 100-3
    distributed, 104-8
    sprite in block, 163-68

Screen saver, 169-74
Sequence detector, 42-47
7-Segment decoder, 20-22
7-Segment displays, 20-22
    multiplexing, 56-59
Spartan-3E, (*see* FPGA)
Sprite, 158-68
Square root (*see* algorithms)
Star, (*see* graphics)

State diagram, 13, 46, 49, 67, 82, 122, 132, 136, 139, 183, 196, 218, 223, 233, 245, 255, 291
State machine, 41-53
    Mealy, 41-42, 45-47
    Moore, 41-45
STD_LOGIC, 5-6
STD_LOGIC_VECTOR, 5-6

Test bench, 204-14, 333-37
Turnkey system, 342-46
Tutorial, (*see* Aldec)

UART, 130-145

Verilog, 2
VHDL, 1-3
    assert statement, 205-6
    case statement, 18
    for loop, 25-41, 61-63
    generic statement, 5-6
    if statement, 4-5
    package, 53-55
    port map statement, 6-7
    process, 4
    quick reference guide, 347-48
    selected signal assignment statement, 22
    sensitivity list, 4
    wait statement, 205
    while statement, 59-61

x7segb, 56-59
Xilinx, 1, 3
    ISE WebPACK, 3
XPower, 338-41

Printed in the United States
213941BV00002B/6/P